METHODS IN MOLECULAR BIOLOGY

Series Editor
John M. Walker
School of Life and Medical Sciences
University of Hertfordshire
Hatfield, Hertfordshire, AL10 9AB, UK

For further volumes:
http://www.springer.com/series/7651

PCR

Methods and Protocols

Edited by

Lucília Domingues

CEB—Centre of Biological Engineering, University of Minho, Braga, Portugal

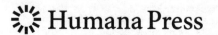 Humana Press

Editor
Lucília Domingues
CEB—Centre of Biological Engineering
University of Minho
Braga, Portugal

ISSN 1064-3745   ISSN 1940-6029   (electronic)
Methods in Molecular Biology
ISBN 978-1-4939-8381-0   ISBN 978-1-4939-7060-5   (eBook)
DOI 10.1007/978-1-4939-7060-5

Printed on acid-free paper

This Springer imprint is published by Springer Nature
The registered company is Springer Science+Business Media LLC
The registered company address is: 233 Spring Street, New York, NY 10013, U.S.A.

Preface

Discovered by Kary Mullis in 1983, the polymerase chain reaction (PCR) is a breakthrough technology that has allowed the advancement of different scientific fields, being a fundamental tool in current scientific research. As such, significant literature exists on the basics of the PCR technique, but the specificities of its application to different areas of the biotechnology and bioengineering field is mostly dispersed. Despite being a well-established technique, novel applications are constantly emerging given the power and flexibility of PCR, with continuous updates being published in this very exciting and moving field. Thus, along with cutting-edge methodologies, this volume focuses on many core PCR applications in the biotechnology and bioengineering field.

In the initial chapters, software for in silico PCR and primer design as well as some particular PCR protocols, such as long fragment or degenerate PCR, are given. Subsequently, particular focus is given to PCR applied to molecular and synthetic biotechnology, including the presentation of a novel platform for high-throughput gene synthesis by PCR. Also, a protocol is presented for nucleic acid extraction and molecular enrichment by whole genome amplification (CRENAME) before PCR, which enables the multiparametric assessment of potable/drinking water, but that can be easily applied to other type of samples. In the following chapters, several examples of PCR applications in food science and technology, environmental microbiology and molecular ecology, and healthcare are presented. Within these, novel applications in currently hot research topics, such as synthetic biology, food authentication and metagenomics are addressed. The book is not intended to cover all existing PCR protocols, but rather to give an overview of the power and flexibility of this technique with concrete examples in the biotechnology and bioengineering field. While initially aimed to cover only end-point PCR, two chapters dealing with the detection of allergens and soy in food matrices include real-time PCR protocols, as in these cases quantification is mandatory.

In the fast-changing world of today, it's hard to imagine how a simple technique not only managed to stay popular for over 30 years, as its use is still expanding. One of the ways PCR has managed to maintain ubiquitous has been through diverse technological advances. This book aims to contribute to a current update of PCR-dependent methods and thus, to be a valuable and useful resource for wet lab researchers, particularly within the biotechnology and bioengineering field.

Braga, Portugal   *Lucília Domingues*

Contents

Contributors

TATIANA Q. AGUIAR • *CEB—Centre of Biological Engineering, University of Minho, Braga, Portugal*

JOANA ISABEL ALVES • *CEB—Centre of Biological Engineering, University of Minho, Braga, Portugal*

MARIA MADALENA ALVES • *CEB—Centre of Biological Engineering, University of Minho, Braga, Portugal*

FLÁVIO AZEVEDO • *Centre of Molecular and Environmental Biology (CBMA), Department of Biology, University of Minho, Braga, Portugal*

PREETI BAJPAI • *Department of Biosciences, Integral University, Lucknow, India*

MÁRIO BARROCA • *Centre of Molecular and Environmental Biology (CBMA), Department of Biology, University of Minho, Braga, Portugal*

MICHEL G. BERGERON • *Département de microbiologie-infectiologie et d'immunologie, Faculté de médecine, Université Laval, Québec City, QC, Canada*

SARITA GANAPATHY BHAT • *Department of Biotechnology, Cochin University of Science and Technology, Cochin, Kerala, India*

LUC BISSONNETTE • *Centre de recherche en infectiologie de l'Université Laval, Axe maladies infectieuses et immunitaires, Centre de recherche du CHU de Québec-Université Laval, Québec City, QC, Canada*

JOANA L.A. BRÁS • *Faculdade de Medicina Veterinária, Centro Interdisciplinar de Investigação em Sanidade Animal (CIISA), Universidade de Lisboa, Lisbon, Portugal; NZYTech Genes & Enzymes, Estrada do Paço do Lumiar, Lisbon, Portugal*

ORIANNA BRETSCHGER • *J. Craig Venter Institute, La Jolla, CA, USA*

LEIGH BURGOYNE • *School of Biological Sciences, Flinders University, Adelaide, South Australia, Australia*

MARIA JORGE CAMPOS • *MARE – Marine and Environmental Sciences Centre, ESTM, Instituto Politécnico de Leiria, Peniche, Portugal*

DAVID CATCHESIDE • *School of Biological Sciences, Flinders University, Adelaide, South Australia, Australia*

ENG WEE CHUA • *Faculty of Pharmacy, Universiti Kebangsaan Malaysia, Kuala Lumpur, Malaysia*

TONY COLLINS • *Centre of Molecular and Environmental Biology (CBMA), Department of Biology, University of Minho, Braga, Portugal*

JOANA T. CUNHA • *CEB—Centre of Biological Engineering, University of Minho, Braga, Portugal*

LUCÍLIA DOMINGUES • *CEB—Centre of Biological Engineering, University of Minho, Braga, Portugal*

MONTSERRAT ESPIÑEIRA • *ANFACO-CECOPESCA, Carretera Colegio Universitario, Vigo, Spain*

VÂNIA O. FERNANDES • *Faculdade de Medicina Veterinária, Centro Interdisciplinar de Investigação em Sanidade Animal (CIISA), Universidade de Lisboa, Lisbon, Portugal; NZYTech Genes & Enzymes, Estrada do Paço do Lumiar, Lisbon, Portugal*

CARLOS M.G.A. FONTES • *Faculdade de Medicina Veterinária, Centro Interdisciplinar de Investigação em Sanidade Animal (CIISA), Universidade de Lisboa, Lisbon, Portugal; NZYTech Genes & Enzymes, Estrada do Paço do Lumiar, Lisbon, Portugal*

CATARINA I.P.D. GUERREIRO • *NZYTech Genes & Enzymes, Estrada do Paço do Lumiar, Lisbon, Portugal*

BJÖRN JOHANSSON • *Centre of Molecular and Environmental Biology (CBMA), Department of Biology, University of Minho, Braga, Portugal*

TINA KOLLANNOOR JOHNY • *Department of Biotechnology, Cochin University of Science and Technology, Cochin, Kerala, India*

RUSLAN KALENDAR • *National Center for Biotechnology, Astana, Kazakhstan; PrimerDigital Ltd, Helsinki, Finland*

MARTIN A. KENNEDY • *Department of Pathology and Carney Centre for Pharmacogenomics, University of Otago, Christchurch, New Zealand*

BEKBOLAT KHASSENOV • *National Center for Biotechnology, Astana, Kazakhstan*

LIN Y. KOH • *School of Biological Sciences, Flinders University, Adelaide, South Australia, Australia*

FELIPE LOMBÓ • *Research Unit "Biotechnology in Nutraceuticals and Bioactive Compounds-BIONUC", Universidad de Oviedo, IUOPA (Instituto Universitario de Oncología del Principado de Asturias), Oviedo, Spain*

SIMRAN MAGGO • *Department of Pathology and Carney Centre for Pharmacogenomics, University of Otago, Christchurch, New Zealand*

ANDRÉE F. MAHEUX • *AFM water Consulting, Québec City, QC, Canada*

FARINA MUJEEB • *Department of Biosciences, Integral University, Lucknow, India*

ALEXANDR MUTERKO • *Laboratory of Plant Molecular Genetics and Cytogenetics, The Federal Research Center Institute of Cytology and Genetics, Novosibirsk, Russia*

CÉLIA NOGUEIRA • *Research & Development Unit, Human Genetics Department, National Institute of Health Dr Ricardo Jorge, Porto, Portugal; Newborn Screening, Metabolism and Genetics Unit, Human Genetics Department, National Institute of Health Dr Ricardo Jorge, Porto, Portugal*

CARLA OLIVEIRA • *CEB—Centre of Biological Engineering, University of Minho, Braga, Portugal*

NEELAM PATHAK • *Department of Biosciences, Integral University, Lucknow, India*

MARIA ALCINA PEREIRA • *CEB—Centre of Biological Engineering, University of Minho, Braga, Portugal*

HUMBERTO PEREIRA • *Centre of Molecular and Environmental Biology (CBMA), Department of Biology, University of Minho, Braga, Portugal*

SUJAL S. PHADKE • *J. Craig Venter Institute, La Jolla, CA, USA*

ALBERTO QUESADA • *Facultad de Veterinaria, Departamento de Bioquímica, Biología Molecular y Genética, Universidad de Extremadura, Extremadura, Spain*

ANDREIA FILIPA SALVADOR • *CEB—Centre of Biological Engineering, University of Minho, Braga, Portugal*

ERLAN M. RAMANCULOV • *National Center for Biotechnology, Astana, Kazakhstan*

FRANCISCO J. SANTACLARA • *ANFACO-CECOPESCA, Carretera Colegio Universitario, Vigo, Spain*

GUSTAVO SANTOS • *Centre of Molecular and Environmental Biology (CBMA), Department of Biology, University of Minho, Braga, Portugal*

ANA FILIPA SEQUEIRA • *Faculdade de Medicina Veterinária, Centro Interdisciplinar de Investigação em Sanidade Animal (CIISA), Universidade de Lisboa, Lisbon, Portugal; NZYTech Genes & Enzymes, Estrada do Paço do Lumiar, Lisbon, Portugal*

MALIKA SHAMEKOVA • *Institute of Plant Biology and Biotechnology, Almaty, Kazakhstan*

DIOGO SILVA • *Centre of Molecular and Environmental Biology (CBMA), Department of Biology, University of Minho, Braga, Portugal*

TIMOFEY V. TSELYKH • *Biochemistry and Developmental Biology, Medicum, University of Helsinki, Helsinki, Finland; Minerva Medical Research Institute, Helsinki, Finland*

SMITA RASTOGI VERMA • *Department of Biotechnology, Delhi Technological University, Delhi, India*

LAURA VILARINHO • *Research & Development Unit, Human Genetics Department, National Institute of Health Dr Ricardo Jorge, Porto, Portugal; Newborn Screening, Metabolism and Genetics Unit, Human Genetics Department, National Institute of Health Dr Ricardo Jorge, Porto, Portugal*

GERMÁN VILLAMIZAR-RODRÍGUEZ • *Research Unit "Biotechnology in Nutraceuticals and Bioactive Compounds-BIONUC", Universidad de Oviedo, IUOPA (Instituto Universitario de Oncología del Principado de Asturias), Oviedo, Spain*

RENAUD VINCENTELLI • *Unité Mixte de Recherche (UMR) 7257, Architecture et Fonction des Macromolécules Biologiques (AFMB), Centre National de la Recherche Scientifique (CNRS) – Aix-Marseille Université, Marseille, Cedex, France*

KABYL ZHAMBAKIN • *Institute of Plant Biology and Biotechnology, Almaty, Kazakhstan*

Chapter 1

In Silico PCR Tools for a Fast Primer, Probe, and Advanced Searching

Ruslan Kalendar, Alexandr Muterko, Malika Shamekova, and Kabyl Zhambakin

Abstract

The polymerase chain reaction (PCR) is fundamental to molecular biology and is the most important practical molecular technique for the research laboratory. The principle of this technique has been further used and applied in plenty of other simple or complex nucleic acid amplification technologies (NAAT). In parallel to laboratory "wet bench" experiments for nucleic acid amplification technologies, in silico or virtual (bioinformatics) approaches have been developed, among which in silico PCR analysis. In silico NAAT analysis is a useful and efficient complementary method to ensure the specificity of primers or probes for an extensive range of PCR applications from homology gene discovery, molecular diagnosis, DNA fingerprinting, and repeat searching. Predicting sensitivity and specificity of primers and probes requires a search to determine whether they match a database with an optimal number of mismatches, similarity, and stability. In the development of in silico bioinformatics tools for nucleic acid amplification technologies, the prospects for the development of new NAAT or similar approaches should be taken into account, including forward-looking and comprehensive analysis that is not limited to only one PCR technique variant. The software FastPCR and the online Java web tool are integrated tools for in silico PCR of linear and circular DNA, multiple primer or probe searches in large or small databases and for advanced search. These tools are suitable for processing of batch files that are essential for automation when working with large amounts of data. The FastPCR software is available for download at http://primerdigital.com/fastpcr.html and the online Java version at http://primerdigital.com/tools/pcr.html.

Key words Polymerase chain reaction, Isothermal amplification of nucleic acids, DNA primers nucleic acid hybridization, Primer binding site, PCR primer and probe analysis, Degenerate PCR, Probe, Genetic engineering tools, DNA fingerprints

1 Introduction

The polymerase chain reaction (PCR) is a nucleic acid amplification method fundamental to molecular biology and is the most important practical molecular technique for the research laboratory. Currently, a variety of thermocycling and isothermal techniques for amplification of nucleic acids exist. Thermocycling techniques use temperature cycling to drive repeated cycles of

Lucília Domingues (ed.), *PCR: Methods and Protocols*, Methods in Molecular Biology, vol. 1620,
DOI 10.1007/978-1-4939-7060-5_1, © Springer Science+Business Media LLC 2017

DNA synthesis to generate large amounts of newly synthesized DNA in proportion to the original amount of template DNA. The template DNA strands are amplified by the repeated cycles of two or three temperature steps including (1) heat denature (denaturation of a double-strand DNA template into single strands), (2) annealing (annealing of primers on a single-strand DNA template), and (3) extension reaction (elongation of primers by DNA polymerase) [1, 2].

A number of isothermal techniques have also been developed that do not rely on thermocycling to drive the amplification reaction. Isothermal techniques, which utilize DNA polymerases with strand-displacement activity, have been used as a nucleic acid amplification method that can obviate the need for the repeated temperature cycles. For example, in loop-mediated isothermal amplification (LAMP) [3], the nucleic acid strands of the template are mixed with oligonucleotide primers, strand displacement-type DNA synthetase, and nucleic acid monomers, and this mixture is held at a constant temperature (in the vicinity of 65 °C) to promote the reaction. Some other techniques are also dependent on the strand displacement activity of certain DNA polymerases, such as strand displacement amplification (SDA) [4], chimeric displacement reaction (CDR), rolling circle amplification (RCR) [5], isothermal chimeric amplification of nucleic acids, smart amplification process (SMAP) [6], transcription-based amplification system (TAS) [7], self-sustained sequence replication reaction (3SR) [8], helicase-dependent amplification (HDA) [9], single primer isothermal amplification (SPIA) [10], and cross-priming amplification (CPA) [2, 11–15].

Other homogeneous techniques for detecting target sequences involve a probe (quantitative PCR) or microarrays that have been modified such that they can be detected during the course of an amplification reaction. TaqMan and Molecular Beacons assays both use a reporter and a quencher dye attached to the probe. TaqMan assay is an example of homogeneous nucleic acid detection of a target sequence technique that employs a modified probe. TaqMan probes hybridize to the target sequence while it is being amplified. The enzyme responsible for amplifying a target sequence also degrades any hybridized probe in its path. Among the technologies developed for target detection and quantification, the most promising one is probably molecular beacons. Conventional molecular beacons are single-stranded oligonucleotide hybridization probes that form a stem-and-loop structure. The loop portion contains the sequence complementary to the target nucleic acid (either DNA or RNA). The stem is formed due to hybridization of the complementary sequence of the 3′ end with the 5′ end. The ends of a molecular beacon are self-complementary and are not designed to hybridize to a target sequence.

Implementation of a PCR reaction requires a pair of different or the same oligonucleotides (primers). Primer design is a critical step in all types of PCR methods to ensure specific and efficient amplification of a target sequence. The first (forward primer) binds to one strand of DNA and the second (reverse primer) binds to the complementary strand. If the pair of sites that match the primers are separated by an appropriate distance, then the DNA fragment between those sites (known as the PCR product or amplicon) is copied by the polymerase, approximately doubling in abundance with each cycle. An implicit assumption is that stable hybridization of a primer with the template is a prerequisite for priming by DNA polymerase. Thus, the correct selection of primers is a key step of the PCR procedure; and the accuracy of in silico calculations of the interaction between primers and DNA template is critical for the prediction of the virtual PCR results.

As most PCR techniques requires two primer molecules to amplify a specific piece of DNA in one reaction, the melting temperatures of both primers need to be very similar in order to allow proper binding of both at a similar hybridization temperature. Ideally, all sequences in the target set would exactly match the primers and be amplified, while no sequence in the background set would match the primers. In other words, primers have to be very specific, in order to only amplify those pieces of DNA that are the target.

The specificity of the oligonucleotides is one of the most important factors for good PCR. Optimal primers should hybridize only to the target sequence, particularly when complex genomic DNA is used as template. Amplification problems can arise due to primers annealing to repetitious sequences (retrotransposons, DNA transposons or tandem repeats) [16]. Alternative product amplification can also occur when primers are complementary to inverted repeats and produce multiple bands. This is unlikely when primers have been designed using specific DNA sequences (unique PCR). Primers complementary to repetitious DNA may produce many nonspecific bands in single-primer amplification and compromise the performance of unique PCRs. However, the generation of inverted repeat sequences is exploited in generic DNA fingerprinting methods. Often, only one primer is used in these PCR reactions, the ends of the products should consist of an inverted repeat complementary sequence of primer.

In eukaryotes, the inter-repeat amplification polymorphism techniques such as inter-retrotransposon amplified polymorphism (IRAP), retrotransposon microsatellite amplification polymorphisms (REMAP) or inter-MITE amplification have exploited the highly abundant dispersed repeats such as the LTRs of retrotransposons and SINE-like sequences [17, 18]. The association of these sequences with each other makes it possible to amplify a series of bands (DNA fingerprints) using primers homologous to these high copy number repeats. One, two, or more primers are used, pointing outwards from an LTR, and therefore amplifying the tract of DNA between two nearby retrotransposons. IRAP can be carried out with a single primer matching either the 5′ or 3′ end of the LTR but oriented away from the LTR itself, or with two primers. The two primers may be from the same retrotransposon element family or may be from different families. The PCR products, and therefore the fingerprint patterns, result from amplification of hundreds to thousands of target sites in the genome. Retrotransposons generally tend to cluster together in "repeat seas" surrounding "genome islands," and may even nest within each other. Hence, the pattern obtained will be related to the TE copy number, insertion pattern, and size of the TE family [19, 20].

The development of the DNA sequencing technology and, in particularly, the emergence of high-throughput sequencing (next generation sequencing) methods have led to progressive accumulation of huge amounts of raw data about the primary structure of genomes. Currently, many prokaryotes and eukaryotes genomes have been sequenced and annotated in databases. For this reason, the use of in silico approaches is widely demanding for extracting useful information from the input data set and for their further processing using virtual tools to prepare and predict experimental results on the planning stage. One of such methods is the virtual PCR. During in silico PCR the set of primers is tested on target location and amplicon size in one or multiple DNA templates. Although the main goal of in silico PCR is the prediction of the expected products during amplification of the DNA template by the specified primer set other associated tasks are frequently required, such as primer or probe searches, target location, and oligonucleotides design and analysis (for example, evaluation of the melting temperature, prediction of secondary structures including G-quadruplexes detection, hairpins, self-dimers, and cross-dimers in primer pairs) [21, 22].

Currently, several web-based methods of in silico PCR have been implemented [21–30]. Electronic PCR is a web server allowing heuristic searches of predefined genomes with up to two mismatches. UCSC in silico PCR (http://genome.ucsc.edu/cgi-bin/hgPcr) is a web server that uses an undocumented algorithm to search a predefined genome [24]. Primer-BLAST [31] is a web server that uses BLAST [26] as the underlying search method. To the best of our knowledge, all available published search algorithms are heuristic and hence can fail to report some valid hits. It should be noted that none of them is available as stand-alone software, except our FastPCR software [21]. Furthermore, the adaptation of some commonly used sequence similarity search methods to in silico PCR is not entirely successful. In fact, BLAST creates local alignments which may fail to cover the full primer sequence and it does not support searching for pairs of queries separated by arbitrary sequence with variable length. Besides, degenerate primers cannot appear in alignment seeds and therefore require special handling to avoid further false negatives. Also, parameters sensitive enough to find an acceptable number of matches will tend to generate very large numbers of false positives. Significant postprocessing is required to identify valid hits. Primer-BLAST uses a word length of 7, so sites must have an identical 7-mer match to a primer (this can result in false negatives). Therefore, if one consider a 15 nt long primer with differences in the fifth and tenth positions at a given site, the longest identical K-mer at that site would be 5 nt, and BLAST would fail to find the match.

The in silico PCR program must also handle degenerate primers or probes, including those with $5'$ or $3'$ tail sequences and single nucleotide polymorphisms (SNPs). Additionally, bisulfite treated DNA, which contains no cytosine other than the methylated cytosine in a CG dinucleotide, or highly degraded and modified DNA from ancient herbarium and mummies may be used as template. Therefore, an additional task that in silico PCR can perform is the identification of multiple binding sites, including mismatched hybridization, by considering the similarity of the primer to targets along the entire primer sequence.

Furthermore, an in silico PCR program must also handle multiplex, nested, or tilling PCR, which are approaches commonly used to amplify several DNA target regions in a single reaction. In silico PCR aims to test NAAT applications specificity, including the target location and amplicon size in one or multiple target genome(s).

Therefore, the use of primers is not limited to PCR nucleic acid amplification but extends to all standard molecular biology methods. These considerations motivate the development of a new, high-throughput, desirable non-heuristic algorithm that has been implemented as a stand-alone software that includes the virtual PCR possibilities. In developing the FastPCR and the online Java web tools, our aim was to create practical, efficient, and easy-to-use software to multiple primers or probes searches for linear and circular DNA sequences and predict amplicons by in silico PCR of large or small local databases (Table 1) [21]. This in silico tool is useful for quick analysis of primers or probes against target sequences, for determining primer location, orientation, efficiency of binding, and calculating their T_m's. It is also useful to validate existing primers, probes, and their combinations. Probable PCR products can be found for linear and circular templates using standard or inverse PCR as well as multiplex PCR.

This chapter describes the FastPCR software as a complete solution for in silico PCR assay. In particular the interface, configuration, and main capabilities of the program are characterized. The basic and advanced algorithms of primer and probe binding sites searches are discussed in detail. As an example, we demonstrate the use of FastPCR for high-throughput in silico PCR assay of large amounts of data. We also provide the quick start guide for immediate application of FastPCR for in silico investigation of genomes during planning of experiments and for prediction of expected results.

Table 1

Summary of the FastPCR software features for in silico PCR facilities

The possibilities of searching that were included in software for virtual PCR
Finds all possible primer pairings, which are predicted for singleplex or multiplexed primers and probes with potential mismatches, located within specified primer and target;
Against whole genome(s) or a list of chromosomes or circular sequences for prediction of probable PCR amplicons and primer location;
Advanced (complex) search, when two or more sequences linked with each other within a certain distance
Using of oligonucleotides with both standard and degenerate bases for specified primer and target
Additional short or long not complementary sequences at the 5'-termini for primers or both termini for probe allowed
The parameters, controllable either by the user or automatically, are primer length longer than 4 nt
The annealing and the melting temperature for oligonucleotide–target duplex calculated with standard and degenerate oligonucleotides, with the possibility of viewing stable guanine mismatches
The result is presented in the form of alignment of the primers with a target, including the location and similarity of matches and probable PCR amplicons with its constituent primers, length of amplicons and their annealing temperatures

2 Software

2.1 Supported Platforms and Dependencies

The online FastPCR software (http://primerdigital.com/tools/pcr.html) is written in Java with NetBeans IDE (Oracle) and requires the Java Runtime Environment (JRE 8) on a computer. It can be used with any operating system (64-bit OS preferred for large chromosome files).

The FastPCR software (http://primerdigital.com/fastpcr.html) is written in Microsoft Visual Studio 6.0 and compiled to an executive file that, after installation, can be used with any version of Microsoft Windows. For Linux and Mac it requires "Wine" (http://www.winehq.org/) as a compatibility layer for running Windows programs, which is a free alternative implementation of the Windows API that also allows the use of the native Windows DLLs (Dynamic Link Library).

Alternatively, FastPCR software for Windows can run in Linux and Mac using VirtualBox (https://www.virtualbox.org/).

VirtualBox allows user to run more than one operating system at a time. This way, the user can run software written for one operating system on another (for example, Windows software on Linux or Mac) without having to reboot to use it.

2.2 Downloading and Installing

The FastPCR software for Windows is available for download at http://primerdigital.com/fastpcr.html. To install the program, save the **FastPCR.msi** file in your computer. Run the **FastPCR.msi** file that starts the installation wizard. Follow all steps suggested by the wizard. This is a "local" software installation and requires administrator rights to install.

The program manual, licence agreement, and files for installation are available on the internet at http://primerdigital.com/fastpcr/ and the YouTube tutorial videos at http://www.youtube.com/user/primerdigital.

The online FastPCR (jPCR) software requires the Java Runtime Environment (http://www.oracle.com/technetwork/java/javase/downloads/). Oracle strongly recommends that all Java users upgrade to the latest Java 8 release.

Before using the online FastPCR software the user needs to add the URL (http://primerdigital.com/) of this application to Exception Site List (https://www.java.com/en/download/faq/exception_sitelist.xml), which is located under the Security tab of the Java Control Panel (http://www.java.com/en/download/help/appsecuritydialogs.xml). The addition of this application URL to this list will allow it to run after presenting some security warnings. By adding the application URL to the Exception list the users can run Rich Internet Applications (RIAs) that would normally be blocked by security checks. The exception site list is managed in the Security tab of the Java Control Panel. The list is shown in the tab. To add, edit, or remove an URL from the list, click Edit Site List:

- Click on the **Edit Site List** button.
- Click the **Add** in the **Exception Site List** window.
- Click in the empty field under **Location** field to enter the URL: http://primerdigital.com/.
- Click **OK** to save the URL that you entered. If you click Cancel, the URLs are not saved.
- Click **Continue** on the **Security Warning** dialog.

In order to enhance security, the certificate revocation checking feature has been enabled by default starting in Java 7. Before Java will attempt to launch a signed application, the associated certificate will be validated to ensure that it has not been revoked by the issuing authority. This feature has been implemented using both Certificate Revocation Lists (CRLs) and Online Certificate Status Protocol (OCSP) mechanisms.

Optionally, the user can download the self-signed certificates file (http://primerdigital.com/j/primerdigital.cer) and import it to "Signer CA" (Certificate Authority) from Java Control Panel.

Finally, the user needs to set "Security Level" to "High" under the Security tab of the Java Control Panel (as it is shown on the picture: http://primerdigital.com/image/primerdigital_certificate_big.png).

Run and download online FastPCR software from desktop computer using Java Web Start (JavaWS) command:

javaws http://primerdigital.com/j/pcr.jnlp

or

javaws http://primerdigital.com/j/pcr2.jnlp

Alternatively, the user can run the software directly from the WEB site: http://primerdigital.com/tools/pcr.html

2.3 Availability

The FastPCR software is available for download at http://primerdigital.com/fastpcr.html and the online version at http://primerdigital.com/tools/pcr.html. The program manual, licence agreement, and files for installation are available on the Internet at http://primerdigital.com/fastpcr/ and the YouTube tutorial videos at http://www.youtube.com/user/primerdigital.

3 Methods

3.1 In Silico PCR Searching Algorithm

For predicting the sensitivity and specificity of a primer pair it is necessary an algorithm that searches for pairs of sites matching the primers with certain differences, where the sites are separated by a certain distance that represents the range of possible amplicon lengths.

An implicit assumption is that stable hybridization of a primer with the template is a prerequisite for priming by DNA polymerase. Binding occurs if a primer is complementary to the nucleotide sequence of the target. The stable binding of the primer to the target is very important for PCR efficiency, being especially important the stability and complementarity in the 3'-termini of primer from where the polymerase will extend. Mismatches at the 3' end of the primers affect target amplification much more than mismatches at the 5' end. A two base mismatch at the 3' end of the primer prevents amplification, while several mismatches at the 5' end of the primer allow amplification, although with reduced amplification efficiency.

Therefore, the FastPCR software pays particular attention to the 3' end portion of the primer and calculates the similarity of the 3' end of the primer to the target (the length is chosen by the user) to determine the stability of the 3'-terminus. A few mismatches may be tolerated, typically at the expense of reduced amplification efficiency. The parameters adopted are based on our experimental data for efficient PCR and are translated into algorithms in order to design combinations of primer pairs for optimal amplification.

Modeling the hybridization of primers to targeted annealing sites is the only way to predict PCR products. The last 10–12 bases at the 3' end of primers are important for binding stability, as single mismatches can reduce PCR efficiency, the effect increasing with the proximity to the 3' end.

The FastPCR is a quick heuristic search algorithm, designed for PCR primers and probes. FastPCR is computationally efficient, achieving effective complexity that can approach linear time in the database size and constant time in (i.e., independent of) the number of queries. This in silico tool is useful for quickly analyzing primers or probes against target sequences, for determining primer location, orientation, and efficiency of binding, and for calculating their melting and annealing temperature. Moreover, there may be other situations where the PCR product formation can occur with a single primer that is complementary to inverted repeats placed near to each other. Besides, the PCR primers may additionally utilize one or more oligonucleotides (known as probe) for detection of the formed PCR product.

Initially, the FastPCR software creates a hash-table for all K-mers (words of fixed length K) to the primer set. The K-mers' length can be equal to 7, 9, or 12 nt, depending on the sensitivity of the search, type of task and length of the primer. For each K-mers, up to one mismatch inside is stipulated. Therefore, the use of long K-mers (= 12 nt) does not result in loss of sensitivity and does not lead to false negatives. In addition, the software allows the use of a primer sequence shorter (down to 4 nt) than the minimum K-mers (= 7 nt). The FastPCR algorithm searches for a match to a forward primer by enumerating all K-mers in a database sequence. The software offers flexible specificity stringency options. The user can specify the number of mismatches that primers may have to unintended targets at the 3' end region and where these mismatches may be present. The default specificity settings are that at least one mismatch exists in the last seven bases of the 3' end of the primer. The parameters adopted are based on our experimental data for efficient PCR and are translated into algorithms in order to design combinations of primer pairs for optimal amplification. The nucleotide sequence of the 3' end region of the probe may not necessarily be completely complementary to the target. For instance, both termini of the probe "molecular beacon" have not complementary regions to the target.

The parameters for quick alignment can be set to allow different degrees of mismatches at the 3' end of the primers: 0–5 mismatches, being the default number of mismatches 2. Furthermore, the program can handle degenerate primer or probe sequences, including those with 5' or 3' tail. The program includes the detection of the alternative hydrogen bonding to Watson–Crick base pairing, as stable guanine mismatches: G·G, G·T and G·A, during the primer binding site searching.

In the sequence analysis by the consistent screening the software does not create a hash table for the analyzed sequence, just runs through the beginning to the end of the whole sequence and analyses potential sites using a hash-table to match forward and reverse primers or probes. During databases screening for each K-mer, the program allows one mismatch at all locations and accepts additional degenerate bases for both target and primer. In the target sequence, the program ignores long gaps, extended unfinished areas with N. For each primer, the match is extended to the full length of the primer or until $>d$ mismatches and the local similarity for the whole primer sequence is calculated. This procedure is guaranteed to find all valid matches.

Finally, having found the matches to the forward primer and to the reverse primer, the amplicon length can not exceed the maximum length specified by the user. Predicted PCR products can be obtained for linear and circular templates using standard or inverse PCR as well as multiplex PCR or bisulfite treated DNA sequence. The software allows simultaneous testing of single primers or a set of primers designed for multiplex target sequences. The program collects information about sites, found by the primer-specific sequence analysis. The analysis reports contain both the list of the hybridization sites and the amplicon details.

3.2 Melting temperature (T_m) calculation

The T_m for short oligonucleotides with normal or degenerate (mixed) nucleotide combinations are calculated in the default setting using nearest neighbor thermodynamic parameters. The T_m is calculated using a formula based on nearest neighbor thermodynamic theory with unified dS, dH, and dG parameters:

$$T_m(°C) = \frac{dH}{dS + R\ln\left(\dfrac{c}{f}\right) + 0.368(L-1)\ln\left(\left[K^+\right]\right)} - 273.15,$$

where dH is enthalpy for helix formation, dS is entropy for helix formation, R is molar gas constant (1.987 cal/K mol), c is the nucleic acid molar concentration (250 pM), $[K^+]$ is salt molar concentration (default value is 50 mM). The $f = 4$ when the two strands are different and $f = 1$ when self-hybridization takes place.

The T_m for mixed bases is calculated by averaging nearest neighbor thermodynamic parameters (enthalpy and entropy values) for each mixed site, and the extinction coefficient is similarly predicted by the averaging nearest neighbor values of mixed sites.

Mismatched pairs can also be taken into account [32, 33]. The melting temperature for in silico PCR experiments with oligonucleotides with mismatches to the target is calculated using values for the thermodynamic parameters for a nucleic acid duplex.

Tools calculate T_m for primer dimers with mismatches for pure, mixed, or modified (inosine, uridine, or locked nucleic acid) bases using averaged nearest neighbor thermodynamic parameters provided for DNA–DNA duplexes [33]. Besides Watson–Crick base pairing, there is a variety of other hydrogen bonding configurations possible such as G/C-quadruplexes or wobble base pairs that the FastPCR software detects [34, 35].

The programme includes the detection of the alternative hydrogen bonding to Watson–Crick base pairing and in silico PCR primer binding site detection. The mismatches stability follows the trend in order of decreasing stability: G-C > A-T > G·G > G·T ≥ G·A > T·T ≥ A·A > T·C ≥ A·C ≥ C·C. Guanine is the universal base, since it forms the strongest base pair and the strongest mismatches. On the other hand, "C" is the most discriminating base, since it forms the strongest pair and the three weakest mismatches. Therefore, the tools are also looking at stable guanine mismatches: G·G, G·T, and G·A [33].

3.3 Calculation of Optimal Annealing Temperature

The optimal annealing temperature (T_a) is the range of temperatures where the efficiency of the PCR amplification is maximal without nonspecific products. The most important values for estimating the T_a is the primer quality, the T_m of the primers and the length of the PCR fragment. Primers with high T_m's (>60 °C) can be used in PCRs with a wide T_a range compared to primers with low T_m's (<50 °C). The optimal annealing temperature for PCR is calculated directly as the value for the primer with the lowest T_m (T_m^{min}):

$$T_a(°C) = T_m^{min} + \ln L,$$

where L is length of the PCR fragment [22].

4 The Interface

4.1 The Program Interface

The software has a user-friendly interface, containing the menu, toolbars, ribbon, and three text editor tabs. Getting started with a basic project in FastPCR software is as easy as opening a new or existing file, copy-paste, or starting to type. The ribbon is designed to help the user to quickly find the commands that are needed to complete a task. Commands are organized in logical groups, which are collected together under tabs. Each ribbon relates to a type of activity, such as "PCR Primer Design," "In silico PCR," or "Primer Test."

There are three independent text editors at different tabs: "Sequences," "Additional sequence(s) or Pre-designed primers (probes) list," and "Result report." The two first text editors are necessary for loading sequences for analysis: the text editor

"General Sequence(s)" editor is designed for working with the project sequences, and the "Additional sequence(s) or pre-designed primers (probes) list" text editor is applied for special and additional sequences, such as pre-designed primers, multiple query sequences or the numbers for input. To save a file from the current text editor, the users must select the file format they want to save the file in (for example, Rich Text Format (.rtf), Excel worksheet (.xls), or text/plain format (.txt)). Furthermore, the FastPCR clipboard allows users to copy and paste operations of text or table from Microsoft Office documents, Excel worksheet or other programs and paste them into another Office document. For this, it is important that the entire target sequences are prepared with the same format.

4.2 Input Data

The sequence data file should be prepared using a text editor (Notepad, WordPad, Word), and stored in the ASCII format as text/plain or Rich Text Format (.rtf). The software accepts a number of different types of input formats and automatically determines the format of the input. To allow this feature there are certain conventions required with regard to the input of identifiers. The formats accepted as input data can be a single or multiple separate DNA sequences in FASTA format, bare sequence, sequence identifiers, tabulated format (two columns from Excel sheet or Word table), EMBL, MEGA, GenBank and MSF, DIALIGN or simple alignment formats, and Blast Queue web alignments result format. The template length is not limited. NCBI Genbank accession records can be retrieved by querying with an accession number or NCBI sequence identifier through the File menu (Fig. 1).

The software allows opening files in several ways:

- The original file can be opened as read-only for editing with text editors.

- The larger files can be opened directly to memory without using text editors.

- Multiple files can be selected within a folder and opened to memory during task execution.

Furthermore, users can type or import from file(s) into "Sequences" or "Additional sequence(s) or pre-designed primers (probes) list" editors. The entire target sequences to be used for in silico PCR search should be pasted in the "Sequences" tab text area and the primers list sequences should be pasted in the "Additional sequence(s) or pre-designed primers (probes) list" tab text area.

Once a sequence file is open, the software displays the information about opened sequences and sequences format. The information status bar shows the amount of sequences, total sequences length (in nucleotides), nucleotide compositions, purine, pyrimidine, and CG% contents.

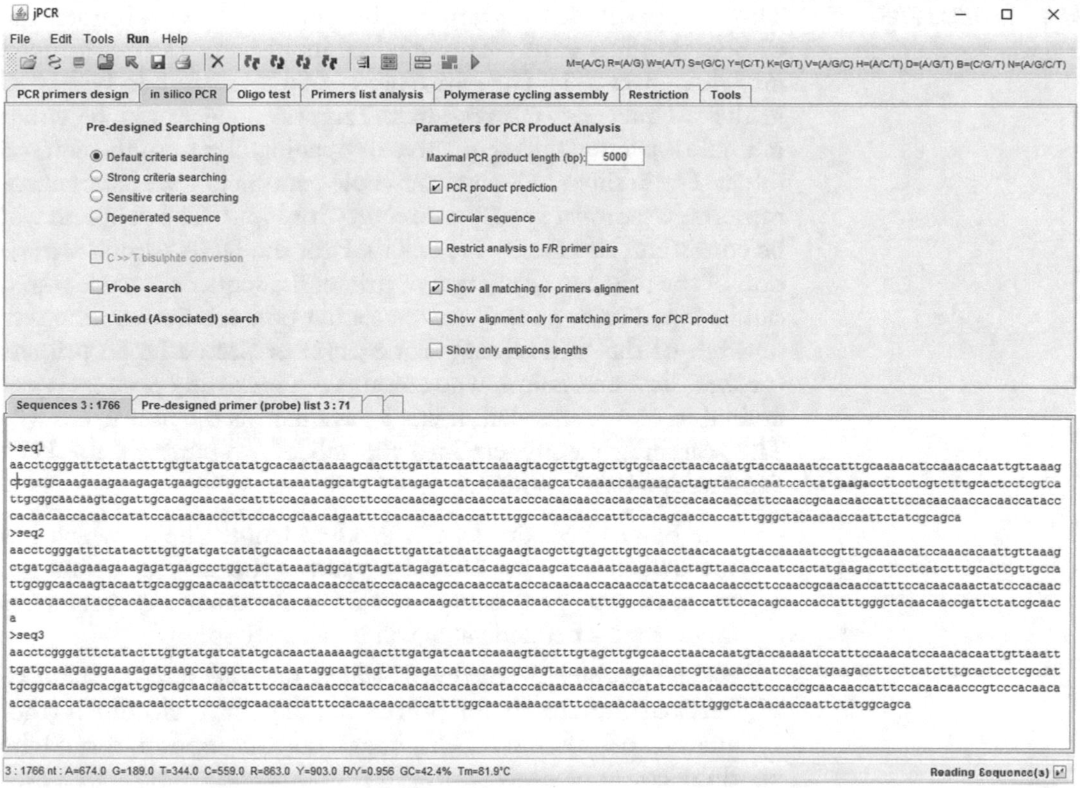

Fig. 1 Input of FastPCR online Java version, primer sequences editor, and user interface

The input DNA sequence can contain degenerate nucleotides accepted as IUPAC code, which is an extended vocabulary of 11 letters that allows the description of ambiguous DNA code [36]. Each letter represents a combination of one or several nucleotides: $M(A/C), R(A/G), W(A/T), S(G/C), Y(C/T), K(G/T), V(A/G/C)$, $H(A/C/T)$, $D(A/G/T)$, $B(C/G/T)$, $N(A/G/C/T)$, and also $U(T)$ and I (Inosine).

4.3 Program Output

The software automatically generates results to the "Result report" text editor in a tabulated format, ready for transferring to a Microsoft Excel sheet with copy-paste. Alternatively, output results can be easily saved as .XLS or .RTF Text files compatible with Excel or Open Office.

The separated output of the primer design comprises a list of primers, a set of primer pair sequences with their theoretical PCR products, and for multiplex PCR, the result of the calculation of multiple PCR primers for given target sequences. The output shows optimal annealing temperature for each primer pair, the size of the PCR product and complete information for each designed primer and for multiplex PCR product set.

4.4 In Silico PCR Application Settings

The in silico PCR program can be initiated by clicking on the ribbon "in silico PCR." The required input items can be grouped into three parts. (1) The entire target sequences should be pasted in the "Sequences" tab text area. Target sequences can be either multiple separate DNA sequences or opening files from the entered folder. For in silico PCR against whole genome(s) or a list of chromosome, user must specify a directory for input. The program will be consistent, and file-by-file will look for the DNA sequence position of the primers. (2) Single or primer list sequences to be tested can be typed in or pasted as pre-existing primer's list into the second tab of the "Additional sequence(s) or pre-designed primers (probes) list" text editor. The amount of preexisting primers is not limited to one primer pair, it can be as much as the user needs. (3) The searching parameters into the tab of "Parameters for PCR Product Analysis" contain:

– The box of "Maximal PCR Product length (bp)," which has the default value of 5000 bp—allows the user to define the maximal size of the expected PCR product; any amplicons larger than a defined value will be filtered out.

– "PCR product prediction," which has the default value of checked—to search for primer binding sites without further analysis of potential PCR fragments this option should be disabled.

– "Circular sequence"—in the analysis of circular molecules (plasmid, mitochondrial or plastid DNA, etc.) the primers can produce one or two amplicons.

– "C >> T bisulphite conversion"—design of specific PCR primers for in silico bisulphite conversion for both strands of bisulphite modified genome sequences; only cytosines not followed by guanidine (CpG methylation) will be replaced by thymines.

– "Restrict analysis for F/R primer pairs"—this option is used for analysis of primer lists, where each primer pairs are united by the common name. Similar analysis can be carried out for primers with the same names or with names that differ in the last letter—F/R. The program will recognize paired primers (Forward as F and Reverse as R). For this type of analysis, primers from the same pair(s) must have identical names and finish with "R or F" (e.g., "seq1R" and "seq1F" form a pair). The name length and structure (including "F" and "R" inside names) are not important. Moreover, the program is not limited in one unique pair per primer: for one "Forward" primer, there can be several "reverse," and the same for the "Reverse" primer. The searching of potential amplicons will be carried out only for these primer pairs, while other primers from list will be ignored.

4.5   Additional Options Relating to Representation of Results

The additional configuration settings allow optimizing the primer or probe binding site search and increasing the representativeness of the results. This is mainly determined by the following parameters:

- "Show all matching for primers alignment," checked by default—the software shows all matches of stable binding primer to the target. Not in all cases the combinations of primers are able to produce the PCR products in the current assay conditions, but the user can examine the stability of primer binding sites, orientation and coordinates in the target.

- "Show alignment only for matching primers for PCR product"—in the previous option all primer binding sites were represented, while in this case the analysis of primer and target alignment will be shown only for matching primers.

- "Show only amplicons lengths"—checking this option allows the user to collect only amplicons' lengths without analysis of primer and target alignment. This option is recommended for in silico PCR of whole genomes, including analysis of all chromosomes with highly abandoned repeated sequences (in silico PCR for techniques based on repeats: iPBS, IRAP, ISSR, or RAPD).

The main search criteria for primer binding sites that the users can select into the "Pre-designed Searching Options" tab are:

- "Default criteria searching"—K-mers = 9 nt with up to a single mismatch at the 3′ termini (single mismatch within K-mers) and at least 15 bases complementarity for primers longer than 14 nucleotides. The "Default criteria searching" is recommended for searching of primer binding sites, because this provides the most effective and fastest searching and minimizes false positives. This criterion is also recommended when using degenerated target DNA or/and primers.

- "Strong criteria searching"—K-mers = 12 nt with a single mismatch at the 3′ termini and at least 15 bases complementarity for primers longer 14 nucleotides. The "Strong criteria searching" is recommended when it is necessary the fast searching of primer binding sites with minimum mismatches at the 3′ termini. Seldom, when primer sequences or target DNA are degenerate and when using the "Strong criteria of searching" no primer binding sites are revealed. In this case, it is necessary to use the "Sensitive criteria searching" and "Degenerated sequence."

- "Sensitive criteria searching"—K-mers = 7 nt with a maximum of two mismatches at the 3′ termini (single mismatch within K-mers) and at least 15 bases complementarity for primers longer than 14 nucleotides.

- "Degenerated sequence"—K-mers = 7 nt with a maximum of three mismatches at the 3′ termini and at least 15 bases complementarity for primers longer than 14 nucleotides.

In some cases, the user can use the option of "Probe search" or "Advanced (Linked) search."

The "Probe search" helps the user to execute searching of binding sites not only for primers but also for probes (TaqMan, molecular beacon, microarrays, etc.). By default, the K-mers value is set equal to 9 with up to a single mismatch within K-mers. However, if the probe length is less than 9 and more than 3, the length of the K-mers will be set equal to the probe's length. This option is recommended when primer binding sites were not found or for searching of binding sites of probes for which the complementarity is expected only for part of the sequence (e.g., in "molecular beacon," where both termini do not have complementary regions to the target).

The "Linked (Associated) search" is a programmable searching that can be used when binding sites for primers or probes are searched within a determined distance. This criterion is described in detail below.

4.6 In Silico PCR Analysis Steps

Running in silico PCR analysis with the FastPCR is a stepwise process. The user has the ability to set all possible parameters required to run in silico PCR analysis interactively:

1. Select a task type at the ribbon—"in silico PCR".

2. Input user data.

   (a) The entire target sequences to be used for in silico PCR search should be pasted in the "Sequences" tab text area.

   (b) The entire primers list sequences should be pasted in the "Additional sequence(s) or pre-designed primers (probes) list" tab text area.

3. Choose an algorithm.

   (a) By default—standard algorithm for matching of stable binding primer to the target searching.

   (b) Probe search.

   (c) Linked search.

4. Set parameters.

5. Run.

6. Visualize results.

---

5 FastPCR as a Complete Solution for In Silico PCR Assay

5.1 Example for Software Validation

To validate the software, the NCBI Eucariota database of genome sequences (https://www.ncbi.nlm.nih.gov/genome/browse/) was used as a target set to measure search performance and accuracy (false negatives and false positives). Primer and probe sequences from the retrotransposons sequence were obtained as query sets.

Primer combinations were extracted by using lengths between 12 and 30 nt for the primers and probes, and a maximal length of 5,000 nt for the "amplicon," i.e., the distance between the start of the forward primer and end of the reverse primer.

The memory required depends only on the analyzed sequence length, because the program loads the entire sequence into RAM. The number of hybridization sites and amplicons produced relate directly to the time needed to perform the simulation (Table 2).

The time and memory have been tested using Intel® Core™ i7-4700HQ (2.4 Ghz) processor machine by running the software using complete genomic DNA sequences of different sizes with a list of primers with standard and degenerate bases, both for specified primers and targets.

The execution time is linearly proportional to the target genome size, from instant (for *Arabidopsis thaliana* genomes) to minutes (for the 3 Gb complete human genome). The RAM required is minimal (4 GB) for the complete human genome.

5.2 In Silico PCR Application: Example 1

For in silico PCR analysis, the Java Web-based application or Windows FastPCR software is suitable, allowing the analysis of a set of sequences. Several simple in silico PCR examples may be found from the menu File: "*in silico* PCR example 1 & 2" and "*in silico* PCR, complex search."

1. Launch the jPCR server through http://primerdigital.com/tools/pcr.html or from the desktop using the Java Web Start (JavaWS) command:

   javaws http://primerdigital.com/j/pcr.jnlp.

Table 2

Search times and total numbers of hits for degenerate primer pair corresponding to highly conserved peptide sequence of plant copia-type reverse transcriptase (RT) for several eukaryotic genomes from EMBL-EBI database

Genome	Size in Mb, (total amount of files)	Time (s)	Number of 270–281 bp amplicons, which were found based on optimal searching criteria with 400 bp limit
Arabidopsis thaliana	126 (9)	10	1
Brachypodium distachyon	275 (6)	17	5
Oryza sativa	380 (12)	25	13
Sorghum bicolor	545 (10)	35	10
Glycine max	965 (20)	60	18
Solanum pennellii	1006 (14)	70	14
Zea mays	2100 (18)	135	22
Homo sapiens	3140 (24)	195	0

2. Prepare the target sequences. The FASTA format is recommended as the most convenient for storage and use of genomic sequences.

For the beginning, the list of query primer(s) should be prepared in FASTA format or as a table (table from Microsoft Office documents or Excel worksheets) or only as bare sequences without spaces between letters:

ITS1	TCCGTAGGTGAACCTGCGG
ITS2	GCTGCGTTCTTCATCGATGC
ITS3	GCATCGATGAAGAACGCAGC
ITS4	TCCTCCGCTTATTGATATGC
ITS5	GGAAGTAAAAGTCGTAACAAGG
KAN2-FP	ACCTACAACAAAGCTCTCATCAACC
KAN2-RP	GCAATGTAACATCAGAGATTTTGAG
L_SP6	TCAAGCTATGCATCCAACGCG
L_T7	TAGGGCGAATTGGGCCCGACG

The first column indicates primers' name, while the second column contains primers' sequence. This is the most convenient format for storage and use of primers in such studies. Within primers' sequence, spaces and no DNA letters are allowed. The primers' name can contain any characters, including only space. Furthermore, the names of the primers can be the identical.

FASTA format, which has a description line starting with the ">" sign followed by a plain DNA sequence, is the widespread format for storage and processing of DNA sequences:

```
>ITS1
TCCGTAGGTGAACCTGCGG
>ITS2
GCTGCGTTCTTCATCGATGC
>ITS3
GCATCGATGAAGAACGCAGC
>ITS4
TCCTCCGCTTATTGATATGC
>ITS5
GGAAGTAAAAGTCGTAACAAGG
>KAN2-FP
ACCTACAACAAAGCTCTCATCAACC
>KAN2-RP
GCAATGTAACATCAGAGATTTTGAG
>L_SP6
TCAAGCTATGCATCCAACGCG
>L_T7
TAGGGCGAATTGGGCCCGACG
```

3. Upload the FASTA file from the selected location, copy and paste, type or import from file(s). The entire target sequences to be used for in silico PCR search should be entered in the "Sequences" tab text area and the primers list sequences should be pasted or loaded from a file in the "Additional sequence(s) or pre-designed primers (probes) list" tab text area. If the target sequences or primers list are recognized by the program, the software will rapidly indicate features for these sequences, such as format, length, CG% content and T_m.

4. Select the "*in silico* PCR" ribbon. Optionally, users can specify searching options: stringency and PCR product detection options. At the stringency options, users can specify the number of mismatches that primers are allowed at the 3′ end. The default specificity settings are for a maximum of two mismatches within the 3′ end of the primer. These mismatches within the 3′ end of the primer should not be located close to each other.

5. To execute the searching task click **F5** or **Run** in the drop-down menu.

5.3 Results Interpretation

Once the in silico PCR analysis is complete, the result will appear in the third **Result** text editor tab, In silico**PCR Result,** when any targets have been found from the designated genome or sequences.

The results in the *In silico* **PCR Result** text editor reports the specificity of the primers (locations, including target position, similarity, and T_m), summary of primer pairs in relation to the PCR template, as well as detailed information on each primer pair, their length and T_a (Fig. 2).

The description line of a primer begins with "*In silico* Primer(s) search for:" followed by the target name and the FASTA description of the target genome sequence. The description line of a template begins with a ">" sign followed by its identification. A query primer begins without a ">" sign followed by its identification, including their original sequence. It will show target-specific primers if found, and the actual targets will be listed along with detailed alignments between primers and targets.

Features of the individual reports of the query primers include representation of their alignments with the target sequence. The actual target fragments will be listed along with detailed alignments and linked query sequences, as well as with detailed information on each query sequence, including locations on target position and its similarity (Fig. 2).

The alignment of query primers to their target template will be shown along with their starting and ending coordinates. Nucleotides on the template which perfectly match with the aligned query are embodied by a vertical bar and those mismatched nucleotides are given as a colon (at least similarity 60%) or by a space.

The products are grouped by the target template where they are found in. One or multiple products can be found with product size

Fig. 2 In silico PCR detailed result for example 1 is shown. The primer-template binding is shown with the presence of mismatches, coordinates on target DNA and its similarity and T_m. It will show target-specific primers if found, the actual targets will be listed along with detailed alignments between primers and targets. If the potential amplicons for primer pairs in relation to the PCR template were found, the detailed information on each primer pair, its length and T_a would be shown

and annealing temperature (T_a), including original primer sequences in FASTA format and position and orientation on the target.

Depending on the task, the user can get full information about all potential primer binding sites within the DNA target and PCR template. For whole genome analysis, a similar approach can be redundant. In this case, the summary of each primer pair and the length of amplicons will be enough.

Results can be saved using the "Save as" function in the menu **File** or can be copied and paste to any text editor, such as **Notepad++** (https://notepad-plus-plus.org/), for further study.

```
In silico Primer(s) search for: 1
1 5'-gcttgtcctcaagcgaaaassa

Position: 251->272     89% Tm = 57.8°C

5-gcttgtcctcaagcgaaaassa->
|||||||||||||||||:::::|
tcgcttgtcctcaagcgarrrnnaagtg

Position: 285->306     86% Tm = 57.8°C

5-gcttgtcctcaagcgaaaassa->
||||||||||||||||:::::::
tcgcttgtcctcaagcgawrwnnratcc

2 5'-cgcagcgttctcataaggtcr

Position: 1074<-1094     95% Tm = 58.5°C

<-rctggaatactcttgcgacgc-5
::||||||||||||||||||||
cgssaccttatgagaacgctgcgacgc

>1 251->272
5'-gcttgtcctcaagcgaaaassa
>2 1074<-1094
5'-cgcagcgttctcataaggtcr
PCR product size: 844bp Ta=66°C

>1 285->306
5'-gcttgtcctcaagcgaaaassa
>2 1074<-1094
5'-cgcagcgttctcataaggtcr
PCR product size: 810bp Ta=66°C
```

5.4 In Silico PCR Application: Example 2

To obtain genome sequence data, the entire genome sequence content of the EMBL-EBI database or any other genome databases source has to be downloaded in a FASTA format into local folders on your computer.

1. Access the EMBL-EBI database through http://www.ebi.ac.uk/genomes/eukaryota.html or collect a plant genome sequence from publicly available databases, such as NCBI (https://www.ncbi.nlm.nih.gov/genome/browse/).

2. Select the download format and, optionally, customize the FASTA header.

3. Click on any file from the folder in the File Menu or select **Working with all files from folder (Ctrl-J)** to enter the entire target genome sequences in the "Sequences" tab without opening to Editor.

4. The degenerate primer pair corresponding to highly conserved peptide sequence of plant copia-type reverse transcriptase (RT):

>**RT+(QMDVK)**

5′-CARATGGAYGTNAARAC

>**RT-(YVDDML)**

5′-CATRTCRTCNACRTA

should be pasted into the "Additional sequence(s) or pre-designed primers (probes) list" tab text area.

5. Select the "*in silico* PCR" ribbon. The default specificity settings are set to a maximum of two mismatches within the 3′ end of the primer. These mismatches should not be located close to each other. Set the "Maximal PCR product length (bp)" equal to 1000 and select the "Show only amplicons lengths."

6. To execute the searching task click **F5** or **Run** in the drop-down menu.

Results from the analysis of the human and of some plant genomes are represented in Table 2. For the human genome, these primers cannot be used due to strong differences in the sequence of the Reverse Transcriptase gene from animals and plants, which explains the absence of amplicons for the human genome.

The analysis' summary information, including amplicons, their length and T_a, will be presented for each file separately (nuclear DNA, plastid and mitochondrial DNA) (Fig. 3). For the present example, the length of the amplicons ranged from 270 to 281 bp, which corresponds to the expected value.

Similar to this example, the in silico PCR can be carried out for whole genome analysis as well as for any other tasks, such as in silico RAPD or other methods, based on using repeat DNA sequences. Also, it is likely to be suitable for genotyping or DNA fingerprinting, by using random primers for retroelements from published genomes.

Furthermore, a similar analysis can be carried out for detecting sequences inserted into plasmid vectors after DNA sequencing. based on the ends of the inserted sequence. For this, instead of using primers, one may use small sequences flanking the insertion or within the insertion. In this case, the "Probe search" option should be applied.

5.5 Programmed (Linked, Associated) Searching

The in silico PCR is one example of sequence similarity searching, in which primer sequences are placed at a certain distance and oriented to each other.

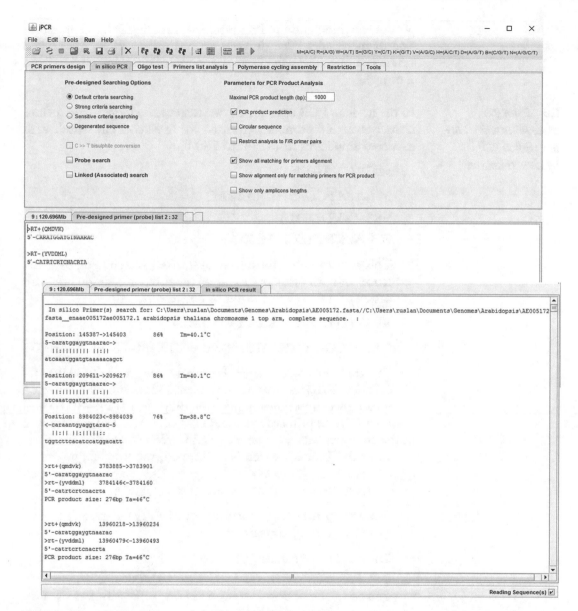

Fig. 3 In silico PCR detailed result for example 2 is shown

Programmed (also known as **Linked** or **Associated**) searching (**LS, Linked search**) allows us to carrying out the advanced searching of primer-template binding sites for any cases, including in silico PCR primers or probes searching. In this kind of analysis, the primer-template binding sites search (for single or more primers or probes) is performed based on a predetermined distance between primer annealing sites. The **Linked search** is able to solve any problems that are associated with primer-template binding sites searching, and the in silico PCR is only one of these tasks. For similar tasks, the only difference between in silico PCR primers or probes searching and the Programmed searching is the representation of the results. Also, in contrast to in silico PCR, the Linked search is unspecified and allows sequences of different length (from 4 nt up).

5.6 Example of an Alternative Way of In Silico PCR by Linked Search

In the in silico PCR example 2, two degenerate Copia-type reverse transcriptase RT primers were used for in silico PCR search with expected amplicons from 200 to 300 bp:

>**RT+(QMDVK)**

5′-CARATGGAYGTNAARAC

>**RT-(YVDDML)**

5′-CATRTCRTCNACRTA

This task can also be solved by Linked search. For this, the sequences of both primers should be rewritten in a single line with indication of the expected distance between them:

>**RT+(QMDVK)_RT-(YVDDML)**

CARATGGAYGTNAARAC [200–300] **TAYGTNGAYGAYATG**

In this example, the Forward primer (5′-CARATG-GAYGTNAARAC) is written on the left, followed by the expected distance between primer-template binding sites ([200–300]) and then the Reverse primer (5′-CATRTCRTCNACRTA) represented in the complementary sequence (**TAYGTNGAYGAYATG**).

After the **Linked search** is performed, the local alignment of query sequences (primers) and target DNA sequence between primer-template binding sites is presented (Fig. 4):

```
atcaaatggatgtcaagtcggccttct/../tgatcgcat-
gcttatatgtagatgacttgat

In silico Primer(s) search for:
C:\Users\Genomes\Arabidopsis\AE005172.
fasta//arabidopsis thaliana chromosome 1 top
arm, complete sequence:

>RT+(QMDVK)_RT-(YVDDML)
CARATGGAYGTNAARAC[200-300]TAYGTNGAYGAYATG

Position: 3783885->3784160      75%
5-CARATGGAYGTNAARAC            TAYGTNGAYGAYATG->
  ||||||||:|| ||: |            ||:|| ||:||| ||
atcaaatggatgtcaagtcggccttct/../
tgatcgcatgcttatatgtagatgacttgat
```

This way, the complete sequence of the target DNA fragments between primer-template binding sites can be obtained, a feature that is unavailable in the in silico PCR. For the presented example,

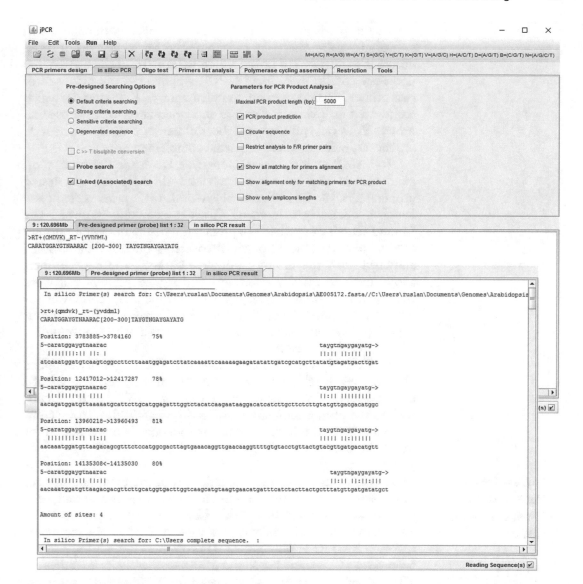

Fig. 4 In silico PCR **Linked search** detailed result for the search of two degenerate Copia-type reverse transcriptase (RT) primers in Arabidopsis genome is shown

the complete sequence of the Arabidopsis reverse transcriptase pseudogene could be obtained, with the primer-template binding sites highlighted in bold:

at**CAAATGGATGTCAAGTC**ggccttcttaaatggagatctt-
gaagaagaagtttacattgagcaaccacaaggctacatagt-
caaaggtgaagaagacaaagtcttgaggctaaaaaaggtgctt-
tatggattaaaacaagccccaagagcttggaatactcgaatt-
gacaagtatttcaaggagaaagatttcatcaagtgtccatat-
gagcatgcactctatatcaaaattcaaaaagaagatatatt-
gatcgcatgctta**TATGTAGATGACTTG**at

In the prediction of possible PCR amplicons obtained with two primer pairs, there are tasks required for when two or more different sequences are linked together. For the in silico PCR, the necessary condition to form a PCR amplicon is the existence of two primer binding sites on complementary DNA strands, which are located at a certain distance and orientation relative to each other. However, primers can be different or equal nucleotide sequences, with complete or partial complementarity.

In LAMP, strand displacement-type DNA amplification, template nucleic acid strands are mixed with three or four nested primer pairs [3, 25]. The analysis of LAMP primers/template match is necessary to control eff icient assay amplification. The **Linked search** can be used to predict the match of existing primer sets and to determine if a newly discovered sequence variant can be amplified with the existing primers. **Linked search** of primers allows users to quickly determine if designed primers will work with genome homologically related sequences, consensus or repeated sequences. Therefore, the LAMP assay with three or four nested primer sets cannot be easily analyzed with conventional PCR simulation software.

In **Linked search**, the complementarity of primer-template binding sites searching criteria are determined by the features of a given region and type of searching (template). Therefore, in contrast to the primer-template binding sites search, the strong complementarity of the 3′-termini of the sequence is not required for the probe-template binding sites searching.

5.7 Linked Search Example: Cassandra LTR Retrotransposon Searching

The in silico searching and extraction from the plant genomes of the LTR-retrotransposon Cassandra sequences is as an example of **Linked search** utilization. The Cassandra retroelement LTR-retrotransposon, universally carries conserved 5S rRNA sequences and associated RNA polymerase III promoters in their long terminal repeats (LTR), and are found in all vascular plants investigated. The use of conservative sequences for all LTR-retrotransposon Cassandra allowed us to find new copies of this retrotransposon in plant genomes, and revealed the presence of the Cassandra LTR-retrotransposon in plant species not previously identified.

5S rRNA sequences from the Cassandra LTR-retrotransposon contains two conservative regions—boxA (RGTTAAGYRHGY) and boxC (RRRATRGGTRACY), separated by 18 nt. Furthermore, in the center of the Cassandra LTR-retrotransposon, it is located a conserved segment, the PBS sequence (TGGTATCAGAGC). Within PBS (primer binding site), located near to the 3′ end of the 5′ LTR, there is a 12–18 bp sequence, complementary to the 3′ tail of some tRNA.

The PBS of the Cassandra LTR-retrotransposon is located in the LTR at 8 bp from the sequence encoding the 5S rRNA for ferns and 173 bp or longer for *Brassica* species. For this reason, the search for the Cassandra LTR-retrotransposon sequence in plant and ferns genomes can be specified like:

>**Cassandra_LTR_PBS**

RGTTAAGYRHGY   [15–25]   RRRATRGGTRACY [5–200] **TGGTATCAGAGC**

To run the Linked search of the Cassandra LTR-retrotransposon:

1. Click on any file from the folder or in the File Menu **Working with all files from folder (Ctrl-J)** to enter the entire target genome sequences in the "Sequences" tab without opening to Editor.

2. The query degenerated sequences (5′ → 3′) corresponding to highly conserved sequence of plant the Cassandra TRIM LTR-retrotransposon in FASTA format:

>**Cassandra_LTR_PBS**

RGTTAAGYRHGY [15–25] RRRATRGGTRACY [5–200] TGGTATCAGAGC

should be pasted into the "Additional sequence(s) or pre-designed primers (probes) list" TAB text area.

3. Select the "*in silico* PCR" ribbon and check "**Linked (Associated) search**", no need to specify any other parameters.

4. To execute the searching task click **F5** or **Run** in the drop-down menu. The output results will be displayed in a new "*In silico* PCR result" tab. The running time for **Linked search** query can last up to a few minutes, depending on the length of the target genome.

Once the **Linked search** analysis is complete, the results will appear in the third **Result** text editor tab, In silico**PCR Result**, if any targets have been found from the designated genome or sequences. Detailed results are reported in the **Linked search** text editor.

The actual target fragments will be listed along with detailed alignments, linked query sequences, as well as detailed information on each query sequence, including locations on the target and similarity (Fig. 5).

The description line for a template begins with "*In silico* Primer(s) search for:", followed by the file name and the FASTA description of the target genome sequence (chromosome number, version of the assembly, and GenBank accession number). A query sequence begins without a ">" sign followed by its identification, including their original sequence.

Fig. 5 In silico PCR **Linked search** detailed result for the example Cassandra LTR retrotransposon searching is shown

The alignment of the query sequence to the target template will be shown along with its starting and ending coordinates. Nucleotides on the template that match perfectly with the aligned query are embodied by a vertical bar and mismatched nucleotides are given as a colon (at least similarity 60%) or by a space.

These result can be saved using the "Save as" function in the menu File or can be copied and paste to any text editor for further data analysis.

```
atagttaagcgtgcttgggctagagtagtttcacgataggt-
gaccttccggga/../gggcgttacaagtggtatcagagccaaa
```

*In silico*Primer(s) search for: AE005173.
fasta//ENA|AE005173|AE005173.1*Arabidopsis
thaliana*chromosome 1 bottom arm, complete
sequence.

```
1 5'-RGTTAAGYRHGY[15-25]RRRATRGGTRACY[5-200]
TGGTATCAGAGC
```

```
Position: 1172475<-1172249     83%
```

```
5-RGTTAAGYRHGY  RRRATRGGTRACY TGGTATCAGAGC->
||||||||: ||    |::||||||:|||    |||||||||||
atagttaagcgtgcttgggctagagtagtttcacgataggt-
gaccttccggga/../gggcgttacaagtggtatcagagccaaa
```

6   Perspectives for Further Development

Like the Linked search example represented above (Cassandra LTR retrotransposon searching) similar investigations can be performed to reveal closely related sequences of genes or genetic elements in eukaryotic and prokaryotic species.

The possibilities of this analysis have a wide application and can be useful to solve many different tasks. The results can be directly used for analysis or for further steps of investigation. The above examples disclose the basic idea, possibilities, potential and limitations of this approach.

For further development, the approach should consider all possible research directions and methodological techniques. For example, the tools for DNA sequence analysis after restriction digestion or nickase treatment should be added.

For this reason, more tools should be developed to increase opportunities of the programmable searching, with addition of functions and subtasks, such as the use of advanced features of language LINQ (Language-Integrated Query) that extend query capabilities, and provide easily learned patterns for querying and updating data. Furthermore, this technology can be extended to support potentially any kind of data stored.

Acknowledgments

Java Web tools are publicly available. They may not be reproduced or distributed for commercial use. This work was supported by the companies Primer Digital Ltd.

References

1. Walker-Daniels J (2012) Current PCR methods. Mat Methods 2:119. doi:10.13070/mm.en.2.119

2. Tisi LC et al. (2010) Nucleic acid amplification. Canada Patent CA2417798

3. Notomi T et al (2000) Loop-mediated isothermal amplification of DNA. Nucleic Acids Res 28(12):e63. doi:10.1093/nar/28.12.e63

4. Walker GT et al (1992) Strand displacement amplification—an isothermal, in vitro DNA amplification technique. Nucleic Acids Res 20(7):1691–1696. doi:10.1093/nar/20.7.1691

5. Banér J et al (1998) Signal amplification of padlock probes by rolling circle replication. Nucleic Acids Res 26(22):5073–5078. doi:10.1093/nar/26.22.5073

6. Tatsumi K et al (2008) Rapid screening assay for KRAS mutations by the modified smart amplification process. J Mol Diagn 10(6):520–526. doi:10.2353/jmoldx.2008.080024

7. Kwoh DY et al (1989) Transcription-based amplification system and detection of amplified human immunodeficiency virus type 1 with a bead-based sandwich hybridization format. Proc Natl Acad Sci U S A 86(4):1173–1177

8. Fahy E et al (1991) Self-sustained sequence replication (3SR): an isothermal transcription-based amplification system alternative to PCR. PCR Methods Appl 1(1):25–33. doi:10.1101/gr.1.1.25

9. Vincent M et al (2004) Helicase-dependent isothermal DNA amplification. EMBO Rep 5(8):795–800. doi:10.1038/sj.embor.7400200

10. Kurn N et al (2005) Novel isothermal, linear nucleic acid amplification systems for highly multiplexed applications. Clin Chem 51(10):1973–1981. doi:10.1373/clinchem.2005.053694

11. Fang R et al (2009) Cross-priming amplification for rapid detection of Mycobacterium tuberculosis in sputum specimens. J Clin Microbiol 47(3):845–847. doi:10.1128/JCM.01528-08

12. Zhao Y et al (2015) Isothermal amplification of nucleic acids. Chem Rev 115(22):12491–12545. doi:10.1021/acs.chemrev.5b00428

13. Katja Niemann VT (2015) Isothermal amplification and quantification of nucleic acids and its use in microsystems. J Nanosci Nanotechnol 06(03). doi:10.4172/2157-7439.1000282

14. Fakruddin M et al (2013) Nucleic acid amplification: alternative methods of polymerase chain reaction. J Pharm Bioallied Sci 5(4):245–252. doi:10.4103/0975-7406.120066

15. Liu W et al (2015) Polymerase spiral reaction (PSR): a novel isothermal nucleic acid amplification method. Sci Rep 5:12723. doi:10.1038/srep12723

16. Smykal P et al (2009) Evolutionary conserved lineage of Angela-family retrotransposons as a genome-wide microsatellite repeat dispersal agent. Heredity (Edinb) 103(2):157–167. doi:10.1038/hdy.2009.45

17. Kalendar R, Schulman AH (2014) Transposon-based tagging: IRAP, REMAP, and iPBS. Methods Mol Biol 1115:233–255. doi:10.1007/978-1-62703-767-9_12

18. Kalendar R et al (2011) Analysis of plant diversity with retrotransposon-based molecular markers. Heredity 106(4):520–530. doi:10.1038/hdy.2010.93

19. Hosid E et al (2012) Diversity of long terminal repeat retrotransposon genome distribution in natural populations of the wild diploid wheat Aegilops speltoides. Genetics 190(1):263–274. doi:10.1534/genetics.111.134643

20. Belyayev A et al (2010) Transposable elements in a marginal plant population: temporal fluctuations provide new insights into genome evolution of wild diploid wheat. Mobile DNA 1(6):1–16. doi:10.1186/1759-8753-1-6

21. Kalendar R et al (2014) FastPCR software for PCR, in silico PCR, and oligonucleotide assembly and analysis. In: Valla S, Lale R (eds) DNA cloning and assembly methods, Methods in molecular biology, vol 1116. Humana, New York, NY, pp 271–302. doi:10.1007/978-1-62703-764-8_18

22. Kalendar R et al (2011) Java web tools for PCR, in silico PCR, and oligonucleotide assembly and analysis. Genomics 98(2):137–144. doi:10.1016/j.ygeno.2011.04.009

23. Lexa M et al (2001) Virtual PCR. Bioinformatics 17(2):192–193. doi:10.1093/bioinformatics/17.2.192

24. Yu B, Zhang C (2011) In silico PCR analysis. Methods Mol Biol 760:91–107. doi:10.1007/978-1-61779-176-5_6

25. Salinas NR, Little DP (2012) Electric LAMP: virtual loop-mediated isothermal AMPlification. ISRN Bioinform 2012:696758. doi:10.5402/2012/696758

26. Johnson M et al (2008) NCBI BLAST: a better web interface. Nucleic Acids Res 36(Web Server issue):5–9. doi:10.1093/nar/gkn201

27. Boutros PC, Okey AB (2004) PUNS: transcriptomic- and genomic-in silico PCR for enhanced primer design. Bioinformatics 20(15):2399–2400. doi:10.1093/bioinformatics/bth257

28. Bikandi J et al (2004) In silico analysis of complete bacterial genomes: PCR, AFLP–PCR and endonuclease restriction. Bioinformatics 20(5):798–799. doi:10.1093/bioinformatics/btg491

29. Rotmistrovsky K et al (2004) A web server for performing electronic PCR. Nucleic Acids Res 32(Suppl 2):W108–W112. doi:10.1093/nar/gkh450

30. Gardner SN, Slezak T (2014) Simulate_PCR for amplicon prediction and annotation from multiplex, degenerate primers and probes. BMC Bioinformatics 15(1):1–6. doi:10.1186/1471-2105-15-237

31. Ye J et al (2012) Primer-BLAST: a tool to design target-specific primers for polymerase chain reaction. BMC Bioinformatics 13:134. doi:10.1186/1471-2105-13-134

32. Peyret N et al (1999) Nearest-neighbor thermodynamics and NMR of DNA sequences with internal A.A, C.C, G.G, and T.T mismatches. Biochemistry 38(12):3468–3477. doi:10.1021/bi9825091

33. SantaLucia J Jr et al (1996) Improved nearest-neighbor parameters for predicting DNA duplex stability. Biochemistry 35(11):3555–3562. doi:10.1021/bi951907q

34. Lane AN et al (2008) Stability and kinetics of G-quadruplex structures. Nucleic Acids Res 36(17):5482–5515. doi:10.1093/nar/gkn517

35. Shing Ho P (1994) The non-B-DNA structure of d(CA/TG)n does not differ from that of Z-DNA. Proc Natl Acad Sci U S A 91(20):9549–9553

36. Nomenclature for incompletely specified bases in nucleic acid sequences (1984) http://www.chem.qmul.ac.uk/iubmb/misc/naseq.html.

Chapter 2

Introduction on Using the FastPCR Software and the Related Java Web Tools for PCR and Oligonucleotide Assembly and Analysis

Ruslan Kalendar, Timofey V. Tselykh, Bekbolat Khassenov, and Erlan M. Ramanculov

Abstract

This chapter introduces the FastPCR software as an integrated tool environment for PCR primer and probe design, which predicts properties of oligonucleotides based on experimental studies of the PCR efficiency. The software provides comprehensive facilities for designing primers for most PCR applications and their combinations. These include the standard PCR as well as the multiplex, long-distance, inverse, real-time, group-specific, unique, overlap extension PCR for multi-fragments assembling cloning and loop-mediated isothermal amplification (LAMP). It also contains a built-in program to design oligonucleotide sets both for long sequence assembly by ligase chain reaction and for design of amplicons that tile across a region(s) of interest. The software calculates the melting temperature for the standard and degenerate oligonucleotides including locked nucleic acid (LNA) and other modifications. It also provides analyses for a set of primers with the prediction of oligonucleotide properties, dimer and G/C-quadruplex detection, linguistic complexity as well as a primer dilution and resuspension calculator. The program consists of various bioinformatical tools for analysis of sequences with the GC or AT skew, CG% and GA% content, and the purine–pyrimidine skew. It also analyzes the linguistic sequence complexity and performs generation of random DNA sequence as well as restriction endonucleases analysis. The program allows to find or create restriction enzyme recognition sites for coding sequences and supports the clustering of sequences. It performs efficient and complete detection of various repeat types with visual display. The FastPCR software allows the sequence file batch processing that is essential for automation. The program is available for download at http://primerdigital.com/fastpcr.html, and its online version is located at http://primerdigital.com/tools/pcr.html.

Key words PCR primer design, Isothermal amplification of nucleic acids, Software probe design, DNA primers, DNA primers nucleic acid hybridization, Degenerate PCR, Tiling arrays, Primer linguistic complexity, Ligase chain reaction

1 Introduction

The polymerase chain reaction (PCR) is a fundamental tool for the molecular biology genetic analyses, and is the most important practical molecular technique for the research laboratory. At the

Lucília Domingues (ed.), *PCR: Methods and Protocols*, Methods in Molecular Biology, vol. 1620, DOI 10.1007/978-1-4939-7060-5_2, © Springer Science+Business Media LLC 2017 time, there is a variety of thermocycling and isothermal techniques used for amplification of nucleic acids. The thermocycling techniques use a temperature cycling to drive the repeated cycles of DNA synthesis to produce large amounts of new DNA being synthesized in proportion to the original amount of a template DNA. The template DNA strands are amplified by the repeated cycles of several temperature steps including (1) heat denature (denaturation of a double-strand DNA template on single strands), (2) annealing (annealing of primers on a single-strand DNA template), and (3) extension reaction (elongation of primers by the DNA polymerase) [1, 2]. The utility of the method is dependent on the identification of unique primer sequences and the PCR-efficient primers design.

The nucleic acid amplification techniques (NAAT) for detection of target sequences use a probe (quantitative PCR) or microarrays specifically modified to be detected during the course of an amplification reaction. For instance, the TaqMan and Molecular Beacons assays both use a reporter and a quencher dye attached to the probe. The TaqMan assay is an example of technique for the homogeneous nucleic acid detection of a target sequence that employs a modified probe. The TaqMan probes hybridize to the target sequence while it is being amplified. The enzyme responsible for amplification of the target sequence also degrades any hybridized probe in its path. Among the technologies developed for target detection and quantification, the most promising is probably the one involving the molecular beacons. The conventional molecular beacons are single-stranded oligonucleotide hybridization probes that form a stem-and-loop structure. The loop part contains the sequence being complementary to the target nucleic acid (either DNA or RNA). The stem part is formed due to hybridization of the complementary sequence of the 3′ end with the 5′ end. The ends of the molecular beacon are self-complementary and are not supposed to hybridize to a target sequence.

Another approach is the use of isothermal techniques for DNA amplification that do not rely on thermocycling to drive the amplification reaction. A number of such techniques have also been developed so far. The isothermal techniques utilize DNA polymerases with strand-displacement activity and are used as a nucleic acid amplification method that can obviate the need for the repeated temperature cycles. For example, to run a reaction in LAMP [3], the mixture of various reagents is held at a constant temperature (in the vicinity of 65 °C) and includes nucleic acid strands of the template, oligonucleotide primers, the strand displacement-type DNA synthetase, and nucleic acid monomers.

Primer design is a critical step in all types of PCR methods to ensure specific and efficient amplification of a target sequence [4–10]. The specificity of the oligonucleotides is one of the most important factors for good PCR, since optimal primers should hybridize only to the target sequence. Particularly it is essential when complex genomic DNA is used as the template. Amplification problems during reaction can arise due to primers annealing to repetitive sequences (retrotransposons, DNA transposons, or tandem repeats) [11]. Alternative product amplification can also occur when primers are complementary to inverted repeats and produce multiple bands. This is unlikely when primers have been designed using specific DNA sequences (unique PCR). Primers complementary to repetitious DNA may produce many nonspecific bands in a single-primer amplification and compromise the performance of unique PCRs. However, the generation of inverted repeat sequences is widely exploited in the generic DNA fingerprinting methods. Often only one primer is used in these PCR reactions, the ends of the PCR products should consist of an inverted repeat complementary to the sequence of the primer.

Therefore, the use of primers is not limited to the PCR nucleic acid amplification but extends to a number of standard molecular biological methods. These considerations motivate the development of a new, high-throughput and stand-alone software that includes the PCR primers design capabilities.

The adaptation of the PCR method for different applications requires development of a new criteria for the PCR primer and probe design to cover approaches such as real-time PCR, group-specific and unique PCR, combinations of multiple primers in multiplex PCR. The criteria must also allow the possibility of the extension PCR for multi-fragments assembling cloning, bisulfite modification assays and of isothermal DNA amplification techniques, as well as a way to design oligonucleotide sets for long sequence assembly by ligase chain reaction, discovery of simple sequence repeats and their amplification as diagnostic markers, TaqMan, molecular beacon, and microarray oligonucleotides [9, 12, 13].

When developing the FastPCR software (Table 1), our aim was to create a practical and easy-to-use tool for the routine manipulation and analysis of sequences for most PCR applications. The adopted parameters are based on our experimental data for efficient PCR and are translated into the algorithms in order to design combinations of primer pairs for the optimal amplification.

This chapter describes the FastPCR software as a complete solution for the PCR primers design. In particular, we describe here the interface, configuration, and main capabilities of the

Table 1

Summary of the FastPCR software for PCR, in silico PCR, and oligonucleotide assembly and analysis

Features
PCR tool provides comprehensive facilities for designing primers for most PCR applications and their combinations:
Standard, multiplex, long distance, inverse, real-time PCR (LUX and self-reporting), group-specific (universal primers for genetically related DNA sequences) or unique (specific primers for each from genetically related DNA sequences), overlap extension PCR (OE-PCR)—multi-fragments assembling cloning and loop-mediated isothermal amplification (LAMP); single primer PCR (design of PCR primers from close located inverted repeat), automatically detecting simple sequence repeat (SSR) loci and direct PCR primer design, amino acid sequence degenerate PCR, polymerase chain assembly (PCA), design amplicons that tile across a region(s) of interest
A long oligonucleotide can be designed for microarray analyses and dual-labeled oligonucleotides for probes such as molecular beacons
PCA or oligonucleotides assembly—created to automate the design oligonucleotide sets for long sequence assembly by PCR
In silico (virtual) PCR or multiple primer or probe searches, or in silico PCR against whole genome(s) or a list of chromosome prediction of probable PCR products, and search for potential mismatching locations of the specified primers or probes
Testing of individual primers, melting temperature calculation for standard and degenerate oligonucleotides including LNA and other modifications
PCR efficiency, linguistic complexity, dimer and G/C-quadruplex detection, dilution and resuspension calculator
Analysis of features of multiple primers simultaneously, including T_m, GC content, linguistic complexity, dimer formation; optimal T_a
Tool for identifying SSR loci by analyzing the low complexity regions of input sequences
Tool for restriction I–II–III types enzymes and homing endonucleases analysis, find or create restriction enzyme recognition sites for coding sequences
Tool for searching for similar sequences (or primers)
Translates nucleotide (DNA/RNA) sequences to the corresponding peptide sequence in all six frames for standard and degenerate DNA and modifications (inosine, uridine)
The program includes various bioinformatics tools for patterns analysis of sequences with GC:(G − C)/(G + C), AT:(A − T)/(A + T), SW:(S − W)/(S + W), MK:(M − K)/(M + K), purine–pyrimidine (R − Y)/(R + Y) skews, CG%, GA% content and purine–pyrimidine skew, the melting temperature, considers linguistic sequence complexity profiles

program. On specific examples, we show how to use the FastPCR software for the high-throughput PCR assay of large amounts of data. The chapter also provides the quick start guide for an immediate application of the FastPCR software for PCR primers design and genome analyses during planning of experiments as well as for prediction of the results.

2 Software

2.1 Supported Platforms and Dependencies

The online version of the FastPCR software (http://primerdigital.com/tools/pcr.html) is written in Java with NetBeans IDE (Oracle) and requires the Java Runtime Environment (JRE 8) on a computer. The program can be used with any operating system (64-bit OS preferred for large chromosome files). The stand-alone version of the FastPCR software (http://primerdigital.com/fast-pcr.html) can be used with any version of Microsoft Windows.

2.2 Downloading and Installing

The online version of the FastPCR software requires the Java Runtime Environment (http://www.oracle.com/technetwork/java/javase/downloads/). The Oracle company strongly recommends that all Java users upgrade to the latest Java 8 release.

The online version FastPCR software users need to add the URL (http://primerdigital.com/) of this application to the Exception Site List (https://www.java.com/en/download/faq/exception_sitelist.xml), which is located under the Security tab of the Java Control Panel (http://www.java.com/en/download/help/appsecuritydialogs.xml). Adding this application URL to the list will allow it to run after presenting some security warnings. Existence of the application URL in the Exception list allows users to run Rich Internet Applications (RIAs) that would normally be blocked by security checks. The exception site list is managed in the Security tab of the Java Control Panel. The list is shown in the tab. To add, edit or remove a URL from the list, use the following:

- Click on the Edit Site List button.
- Click the Add in the Exception Site List window.
- Click in the empty field under Location field to enter the URL: http://primerdigital.com/.
- Click OK to save the URL that you entered. If you click Cancel, the URLs will not be saved.
- Click Continue in the Security Warning dialog.

In order to enhance security, the certificate revocation checking feature has been enabled by default (starting from Java 7). Before Java attempts to launch a signed application, the associated certificate will be validated to ensure that it has not been revoked by the issuing authority. This feature has been implemented using both Certificate Revocation Lists (CRLs) and Online Certificate Status Protocol (OCSP) mechanisms.

Optionally, users can download self-signed certificates file (http://primerdigital.com/j/primerdigital.cer) and import it to "Signer CA" (Certificate Authority) from the Java Control Panel.

Finally, users need to set "Security Level" to "High" under the Security tab of the Java Control Panel (as it is shown on the picture: http://primerdigital.com/image/primerdigital_certificate_ big.png).

Running and downloading online jPCR software from a desktop computer using the Java Web Start (JavaWS) command:

javaws http://primerdigital.com/j/pcr.jnlp

or

javaws http://primerdigital.com/j/pcr2.jnlp

Alternatively, users can run the software directly from the WEB site: http://primerdigital.com/tools/pcr.html.

2.3   Availability

The FastPCR software is available for download at http://primerdigital.com/fastpcr.html and online version at http://primerdigital.com/tools/pcr.html. Web tools are available here: http://primerdigital.com/tools/. The program manual, license agreement, and files for installation are available on the Internet at http://primerdigital.com/fastpcr/ and YouTube tutorial videos at http://www.youtube.com/user/primerdigital.

3   The Interface

3.1   Inputs to FastPCR

The software contains the menus, the toolbars, the ribbon, and three text editors. The ribbon is designed to help the user to quickly find the commands that are needed to complete a task. Commands are organized in the logical groups, which are displayed together under tabs (Fig. 1). Each tab relates to a type of activity, such as "PCR Primer Design," "in silico PCR," or "Oligo Test."

Getting started with a basic project in the FastPCR software is as easy as opening a new or existing file as well as a copy-paste or starting to type.

There are three independent text editors at different tabs: "General Sequence(s)," "Additional sequence(s) or pre-designed primers (probes) list," and "Result report."

The two first text editors are necessary for loading sequences for analysis, the "General Sequence(s)" text editor is designed for working with the project sequences; the "Additional sequence(s) or pre-designed primers (probes) list" text editor is applied for special and additional sequences, for example, for predesigned primers, multiple query sequences or for the numbers for input.

3.2   Program Output

The FastPCR software automatically generates results in the third text editor named as "Result report." It is performed in a tabulated format for transferring the results to Microsoft Excel sheet from a clipboard with a copy-paste method or to save them as .XLS or .RTF text file, compatible with both MS Excel and Open Office. Moreover, the program also produces results containing the list of primers, a set of primer pair sequences with their theoretical PCR

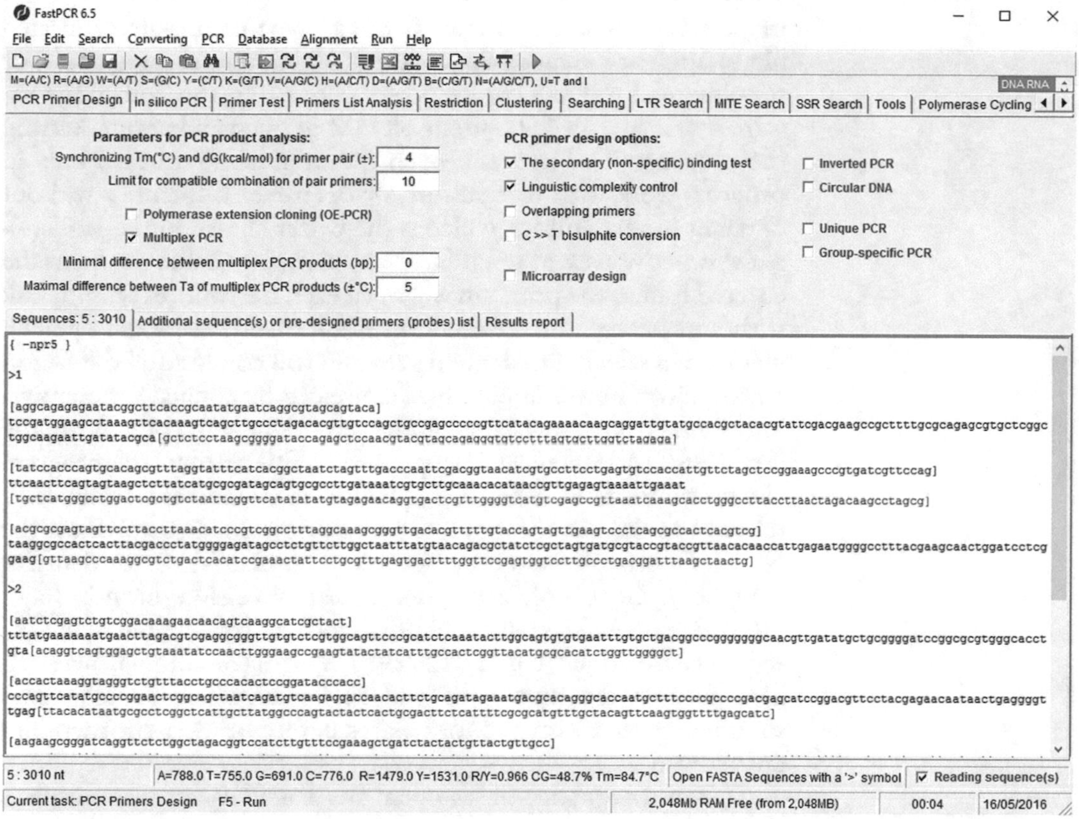

FastPCR 6.5

File Edit Search Converting PCR Database Alignment Run Help

M=(A/C) R=(A/G) W=(A/T) S=(G/C) Y=(C/T) K=(G/T) V=(A/G/C) H=(A/C/T) D=(A/G/T) B=(C/G/T) N=(A/G/C/T), U=T and I   DNA.RNA

PCR Primer Design | in silico PCR | Primer Test | Primers List Analysis | Restriction | Clustering | Searching | LTR Search | MITE Search | SSR Search | Tools | Polymerase Cycling ◀ ▶

Parameters for PCR product analysis:

Synchronizing Tm(°C) and dG(kcal/mol) for primer pair (±): 4
Limit for compatible combination of pair primers: 10

☐ Polymerase extension cloning (OE-PCR)
☑ Multiplex PCR

Minimal difference between multiplex PCR products (bp): 0
Maximal difference between Ta of multiplex PCR products (±°C): 5

PCR primer design options:

☑ The secondary (non-specific) binding test
☑ Linguistic complexity control
☐ Overlapping primers
☐ C >> T bisulphite conversion
☐ Microarray design

☐ Inverted PCR
☐ Circular DNA
☐ Unique PCR
☐ Group-specific PCR

Sequences: 5 : 3010 | Additional sequence(s) or pre-designed primers (probes) list | Results report |

```
{ -npr5 }

>1

[aggcagagagaatacggcttcaccgcaatatgaatcaggcgtagcagtaca]
tccgatggaagcctaaagttcacaaagtcagcttgccctagacacgttgtccagctgccgagccccgttcatacagaaaacaagcaggattgtatgccacgctacacgtattcgacgaagccgcttttgcgcagagcgtgctcggc
tggcaagaattgatatacgca[gctctcctaagcggggataccagagctccaacgtacgtagcagagggtgtccttagtacttagtctagaga]

[tatccacccagtgcacagcgtttaggtatttcatcacggctaatctagtttgacccaattcgacggtaacatcgtgccttcctgagtgtccaccattgttctagctccggaaagcccgtgatcgttccag]
ttcaacttcagtagtaagctcttatcatcgcgcgatagcagtcgacgtccttgataaatcgtgcttgcaaacacataaccgttgagagtaaaattgaaat
[tgctcatgggcctggactcgctgtctaattcggttcatatatatgtagagaacaggtgactcgtttggggtcatgtcgagccgttaaatcaaagcacctgggcctttaccttaactagacaagcctagcg]

[acgcgcgagtagttccttaccttaaacatcccgtcggcctttaggcaaagcgggttgacacgtttttgtaccagtagttgaagtccctcagcgccactcacgtcg]
taaggcgccactcacttacgaccctatgggagatagcctctgttcttggctaatttatgtaacagacgctatctcgctagtgatgcgtaccgtaccgttaacacaaccattgagaatggggcctttacgaagcaactggatcctcg
gaag[gttaagcccaaaggcgtctgactcacatccgaaactattcctgcgtttgagtgacttttggttccgagtggtccttgccctgacggatttaagctaactg]

>2

[aatctcgagtctgtcggacaaagaacaacagtcaaggcatcgctact]
tttatgaaaaaaatgaacttagacgtcgaggcgggttgtgtctcgtggcagttcccgcatctcaaatacttggcagtgtgtgaaatttgtgctgacggcccgggggggcaacgttgatatgctgcggggatccggcgcgtgggcacct
gta[acaggtcagtggagctgtaaatatccaacttgggaaagccgaagtatactatcatttgccactccggttacatgcgcacatctggttggggct]

[accactaaaggtagggtctgtttacctgcccacactccggataccccacc]
cccagttcatatgccccggaactcggcagctaatcagatgtcaagaggaccaacacttcgtcgacatagaaatgacgcacagggcaccaatgctttccccgccccgacgagcatccggacgttcctacgagaacaataactgaggggt
tgag[ttacacataatgcgcctcggctcattgcttatggcccagtactactcactgcgacttctcattttcgcgcatgtttgctacagttcaagtggttttgagcatc]

[aagaagcgggatcaggttctctggctagacggtccatcttgtttccgaaagctgatctactactgttactgttgcc]
```

5 : 3010 nt   A=788.0 T=755.0 G=691.0 C=776.0 R=1479.0 Y=1531.0 R/Y=0.966 CG=48.7% Tm=84.7°C   Open FASTA Sequences with a '>' symbol   ☑ Reading sequence(s)

Current task: PCR Primers Design   F5 - Run   2,048Mb RAM Free (from 2,048MB)   00:04   16/05/2016

Fig. 1 The FastPCR sequence editor and user interface

products and for multiplex PCR. It also provides a user with the results of the calculation of multiple PCR primers for a given target sequences. In addition, the output shows the optimal annealing temperature for each primer pair as well as the size of PCR product and complete information for each primer designed and for the multiplex PCR product set.

3.3 Sequence Entry

Prepare your sequence data file using a text editor (Notepad, WordPad, Word), and save in a ASCII text/plain format or a Rich Text Format (.rtf). The program takes either a single sequence or accepts multiple separate DNA sequences in the FASTA, tabulated (two columns from either MS Excel sheet or Word table), EMBL, MEGA, GenBank and MSF, DIALIGN format or in the simple alignment and Blast Queue web alignments result formats. The template length is not limited.

The FastPCR software clipboard allows user to copy/paste operations with text or table from Microsoft Office documents or Excel worksheets (or other programs) and use them in another MS Office document. Importantly, the full target sequences must be prepared in the same format. User can type or import data from file(s) into the "General Sequence(s)," "Additional sequence(s) or pre-designed primers (probes) list" editors. In the FastPCR software users have several options on how to open a file while starting the program. The user can open the original file as read-only in order to work with text editors, or open file to memory without opening to text editors, which is the better choice for large file(s). An alternative way to open files is by showing to the program the entered folder; the program will open each file while executing task without opening it to text editor. Additionally, users can open all files from a selected folder and program will combine all the files in a text editor. For example, this feature can be applied for converting all files from the selected folder into a single file presenting the list of FASTA sequences. As opposed to this feature, the program allows splitting of FASTA sequences into individual files in a selected folder. At the time, users can download file(s) from the NCBI Genebank server by accession number(s). The identifier may be a Genebank accession, accession.version or gi's (e.g., p01013, AAA68881.1, 129295) and a bar-separated NCBI sequence identifier (e.g., gi|129295). Spaces (or comma) between identifiers in the input will lead to downloading all sequences simultaneously to text-editor (spaces before or after the identifier are allowed).

When a sequence file is open, the FastPCR software displays the information about the opened sequences and the sequences format. The information status bar shows the amount of sequences, total sequences length (in nucleotides), nucleotide compositions, purine, pyrimidine, and CG% contents.

When users save a file from the current text editor, they must choose the file format to save the file, e.g., Rich Text Format (.rtf), MS Excel worksheet (.xls), or text/plain format (.txt).

3.4 FASTA Format Description

The FastPCR software normally is expected to read files with sequences in FASTA format (http://blast.ncbi.nlm.nih.gov/blastcgihelp.shtml). The FASTA format have a highest priority and is simple as the raw sequence proceeded by definition line. The definition line begins with a ">" sign that can be optionally followed by a sequence name of any length and amount of words with no space in between. There can be many sequences listed in the same file. The format requires that a new sequence always starts with a new ">" symbol. It is important to press "Enter" key at the end of each definition line to help the FastPCR software recognize the end and beginning of sequence and the sequence name. Make sure the first line starts with a ">" and, optionally, a header description.

Degenerate DNA sequences are accepted as IUPAC code that is an extended vocabulary of 11 letters that allows the description of ambiguous DNA code [14]. Each letter represents a combination of one or several nucleotides: M = (A/C), R = (A/G), W = (A/T), S = (G/C), Y = (C/T), K = (G/T), V = (A/G/C), H = (A/C/T), D = (A/G/T), B = (C/G/T), N = (A/G/C/T), U = T and I (Inosine).

The program-accepted amino acid codes: A(Ala), C(Cys), D(Asp), E(Glu), F(Phe), G(Cly), H(His), I(Ile), K(Lys), L(Leu), M(Met), N(Asn), P(Pro), Q(Gln), R(Arg), S(Ser), T(Thr), U(Sec), V(Val), W(Trp), Y(Tyr).

3.5 Alignment Format Description

There are many different programs, which produce many different types of alignment formats. The use of a standard set of formats enables creation of programs that can read the results originating from many different sources. In all alignment formats, gaps that have been introduced into the sequences to make them align are indicated by the "-" character. The exception to this rule is GCG/MSF format, which uses "." as the gap character inside the sequences. The alignment file may begin with as many lines of comment or description as required. The first mandatory line that is recognized as part of the MSF file contains the text MSF, or contains the text Alignment as simple alignment format, or contains the texts DIALIGN or MEGA recognized as alignments from these programs. There then follows lines with each sequence line starting with the sequence name which is separated from the aligned sequence residues by white space.

4 The PCR Primers or Probe Design Analysis Options

4.1 PCR Primer Design Generalities

Primer design is one of the key steps for successful PCR. For PCR applications, primers are usually 18–35 bases in length and should be designed such that they have complete sequence identity to the desired target fragment to be amplified. The primer design parameters, controllable either by the user or automatically, include the primer length (12–500 nt), melting temperature for short primers calculated by nearest neighbor thermodynamic parameters, the theoretical primer PCR efficiency (quality at %) value, primer CG content, 3' end terminal enforcement, preferable 3' terminal nucleotide sequence composition in degenerated formulae, and added sequence tags at 5' termini. The other main parameters used for primer selection represented by the general nucleotide structure of the primer such as linguistic complexity (nucleotide arrangement and composition), primer specificity, the melting temperature of the whole primer and the melting temperature at the 3' and 5' termini as well as the self-complementarity and secondary (nonspecific) binding.

The software can dynamically optimize the best primer length for entered parameters. All PCR primer (probe) design parameters are flexible and changeable according to the features of the

Table 2

Default primer design selection criteria

Criteria	Default	Ideal
length (nt)	18–22	>21
T_m range (°C)[a]	49–60	60–68
T_m[a] 12 bases at 3′ end	30–50	40–45
GC (%)	40–70	45–55
3′-end composition (5′-nnn-3′)	swh ssw wsh	ssa, sws, wss
Sequence linguistic complexity (LC, %)[b] ≥70		>90
Sequence Quality (PQ, %)	≥75	>80

[a]Nearest neighbor thermodynamic parameters SantaLucia [16]
[b]Sequence linguistic complexity measurement was performed using the alphabet-capacity l-gram method

analyzed sequence and research task. Primer pairs are analyzed for cross-hybridization, specificity of both primers and, optionally, are selected with similar melting temperatures. Primers with balanced melting temperatures (within 1–6 °C of each other) are desirable but not mandatory. The default primer design selection criteria are shown in Table 2. It is possible to use predesigned primers or probes or, alternatively, predesigned primers can act as references for the design of new primers. The program accepts a list of predesigned oligonucleotide sequences and checks the compatibility of each primer with a newly designed primer or probe.

4.2 Melting Temperature (T_m) Calculation

The T_m is defined as the temperature at which half the DNA strands are in the double-helical state and half are in the "random-coil" state. The T_m for short oligonucleotides with normal or degenerate (mixed) nucleotide combinations are calculated in the default setting using nearest neighbor thermodynamic parameters [15, 16]. The T_m is calculated using a formula based on nearest neighbor thermodynamic theory with unified dS, dH and dG parameters:

$$T_m\left(^{\circ}C\right) = \frac{dH}{dS + R\ln\left(\dfrac{c}{f}\right) + 0.368(L-1)\ln\left(\left[K^+\right]\right)} - 273.15,$$

where dH is enthalpy for helix formation, dS is entropy for helix formation, R is molar gas constant (1.987 cal/K mol), c is the nucleic acid molar concentration (250 pM), [K⁺] is salt molar concentration (default value is 50 mM). The f = 4 when the two strands are different and f = 1 when self-hybridization takes place.

The CG content of an oligonucleotide is the most important factor that influences the T_m value. The melting temperature for mixed bases is calculated by averaging nearest neighbor thermodynamic parameters—enthalpy and entropy values—at each mixed site; extinction coefficient is similarly predicted by averaging nearest neighbor values at mixed sites [5, 6]. Mismatched pairs can be taken into account since the parameters provide for DNA–DNA duplexes and the dangling ends, the unmatched terminal nucleotides [17–19]. The melting temperature for primer (probe) self- or cross-dimers and for in silico PCR experiments with oligonucleotides with mismatches to the target sequence is calculated using values for the thermodynamic parameters for a nucleic acid duplex.

The FastPCR software allows the choice of other nearest neighbor thermodynamic parameters or simple nonthermodynamic T_m calculation formulae. For nonthermodynamic T_m calculation, we suggest using simple formulae; the Wallace–Ikatura rule [20] is often used as a rule of thumb when primer T_m is to be estimated at the bench. However, the formula was originally applied to the hybridization of probes in 1 M NaCl and is an estimate of the melting temperature, for shorter 15 bases:

$$T_m\left(^\circ C\right) = 2\left(L + G + C\right)$$

Alternative and more advanced nonthermodynamic formulae.

$$T_m\left(^\circ C\right) = 64.9 + \frac{41\left(\left[G + C\right] - 16.4\right)}{L}.$$

or formulae [21]:

$$T_m\left(^\circ C\right) = 77.1 + 11.7\log_{10}\left[K^+\right] + \frac{41\left[G + C\right] - 528}{L}$$

where L is the length of primer and $\left[G + C\right]$ is the number of G's and C's, $\left[K+\right]$ is salt molar concentration (default value is 50 mM).

The two equations above assume that the stabilizing effects of cations are the same on all base pairs. Alternatively, the melting temperature of the PCR product may be calculated using the formulae [18]:

$$T_m\left(^\circ C\right) = 81.5 + 16.6\log_{10}\left[K^+\right] + \frac{41\left[G + C\right] - 675}{L}.$$

4.3 Linguistic Complexity of Sequence and Nucleotide-Skew Analysis

The sequence analysis complexity calculation method can be used to search for conserved regions between compared sequences for the detection of low-complexity regions including simple sequence repeats, imperfect direct or inverted repeats, polypurine and polypyrimidine triple-stranded DNA structures, and four-stranded structures (such as G/C-quadruplexes).

Linguistic complexity measurements are performed using the alphabet-capacity L-gram method [22, 23] along the whole sequence length and calculated as the sum of the observed range (xi) from 1 to L size words in the sequence divided by the sum of the expected (E) value for this sequence length.

Linguistic complexity (LC) values for sequence length (s) are converted to percentages, 100% being the highest level:

$$LC(\%) = \frac{100 \times \sum_{L}^{i=1} x_i}{E}, where$$

$$E = \sum_{L}^{i=1} \{ \frac{s-i+1, s < 4^i - 1 + i}{4^i, s \geq 4^i - 1 + i}$$

$$L = \lceil \log_4 \left(\frac{s}{3} \right) + 1 \rceil$$

For example, the sequence 5′-$GTGTGTGTGTGTGTGC$, 17 nt ($L = 3$), contains three nucleotides (G, T, and C), but expected $E = 4$ variants; three variants of two-nucleotides (GT, TG, and GC), but expected $E = (16 - 1)$ variants; three variants of three-nucleotides (GTG, TGT, and TGC), and expected $E = (16 - 2)$ variants. The complexity value is LC = 100 × (3 + 3 + 3) / (4 + 16 − 1 + 16 − 2) = 27.3%.

4.4 Primer Quality (Virtual PCR Efficiency) Determination

Our experimental data showed that the primer nucleotide composition and melting temperature of the 12 bases at the 3′ end of the primers are important factors for PCR efficiency. The melting temperature of the 12 base 3′ terminus is calculated preferably by nearest-neighbor thermodynamic parameters [16]. The composition of the sequence at the 3′ terminus is important; primers with two terminal C/G bases are recommended for increased PCR efficiency [24]. Nucleotide residues C and G form a strong pairing structure in the duplex DNA strands. Stability at the 3′ end in primer template complexes will improve the polymerization efficiency.

We specify an abstract parameter called Primer Quality (PQ) that can help to estimate the efficiency of primers for PCR. PQ is calculated by the consecutive summation of the points according to the following parameters: total sequence and purine–pyrimidine sequence complexity, the melting temperature of the whole primer and of the terminal 3′ and 5′ 12 bases. Self-complementarity, which gives rise to possible dimer and hairpin structures, reduces the final value. PQ tries to describe the likelihood of PCR success of each primer; this value varies from 100 for the best to 0 for the worst primers.

To meet multiplexing demands, it is possible in the program to select the best primer with an optimal temperature range, allowing the design of qualified primers or probes for any target sequence with any CG and repeat content. PQ values of 80 and higher allow for the rapid choice of the best PCR primer pair combination. No adverse effects, due to the modification of the reaction buffer, chosen thermostable polymerases, or variations in annealing temperature, have been observed on the reproducibility of PCR amplification using primers with high PQ.

4.5 Hairpin (Loop) and Dimer Formation

Primer-dimers involving one or two sequences may occur in a PCR reaction. The FastPCR tool eliminates intra- and inter-oligonucleotide reactions before generating a primer list and primer pair candidates. It is very important for PCR efficiency that the production of stable and inhibitory dimers are avoided, especially avoiding complementarity in the 3′-ends of primers from where the polymerase will extend. Stable primer dimer formation is very effective at inhibiting PCR since the dimers formed are amplified efficiently and compete with the intended target.

Primer dimer prediction is based on analysis of non-gap local alignment and the stability of both the 3′ end and the central part of the primers. Primers will be rejected when they have the potential to form stable dimers depending on nucleotides composition and with at least five bases at the 3′ end or seven bases at the central part. Tools calculate T_m for primer dimers with mismatches for pure, mixed, or modified (inosine, uridine, or locked nucleic acid) bases using averaged nearest neighbor thermodynamic parameters provided for DNA–DNA duplexes [15–17, 25, 26].

Besides Watson–Crick base pairing, there is a variety of other hydrogen bonding possible configurations [27–30] such as G/C-quadruplexes or wobble base pairs that the FastPCR software detects.

The program includes the detection of the alternative hydrogen bonding to Watson–Crick base pairing at the primer-dimers and in silico PCR primer binding site detection. The mismatches stability follows the trend in order of decreasing stability: **G·C > A·T > G·G > G·T ≥ G·A > T·T ≥ A·A > T·C ≥ A·C ≥ C·C**. Guanine is the most universal base, since it forms the strongest base pair and the strongest mismatches. On the other hand, "C" is the most discriminating base, since it forms the strongest pair and the three weakest mismatches [25, 31].

Therefore, the tools are also looking at stable guanine mismatches: G·G, G·T, and G·A.

G-rich (and C-rich) nucleic acid sequences can fold into four-stranded DNA structures that contain stacks of G-quartets [30]. These quadruplexes can be formed by the intermolecular association of two or four DNA molecules, dimerization of sequences that contain two G-bases, or by the intermolecular folding of a single strand containing four blocks of guanines; these are easy to eliminate from primer design because of their low linguistic complexity, LC = 32%

for $(\text{TTAGGG})_4$. The software predicts the presence of putative G- and C-quadruplexes in primer sequences. Intermolecular G-quadruplex-forming sequences are detected according to the formula $...G_{m1}X_nG_{m2}...$, where m is the number of G residues in each G-tract (m_1, $m_2 \geq 3$); the gap Xn ($n \leq 2 \times$ minimal ($m_1{:}m_2$)) can be any combination of residues, including G [19]. The gap sequences (Xn) may have varying lengths, and a relatively stable quadruplex structure may still be formed with a loop more than seven bases long, but in general, increasing the length of the gap leads to a decrease in structure stability. It is also possible for one of the gaps to be zero length when there are long poly-G tracts of >6 bases.

4.6 Calculation of Optimal Annealing Temperature

The optimal annealing temperature (T_a) is the range of temperatures where efficiency of PCR amplification is maximal without nonspecific products. The most important values for estimating the T_a is the primer quality, the T_m of the primers and the length of PCR fragment. Primers with high T_m's (>60 °C) can be used in PCRs with a wide T_a range compared to primers with low T_m's (<50 °C). The optimal annealing temperature for PCR is calculated directly as the value for the primer with the lowest T_m (T_m^{min}). However, PCR can work in temperatures up to 10 °C higher than the T_m of the primer to favor primer target duplex formation:

$$T_a\left(^{\circ}\text{C}\right) = T_m^{min} + \ln L,$$

where L is length of PCR fragment.

4.7 The Secondary Nonspecific Binding Test; Alternative Amplification

The specificity of the oligonucleotides is one of the most important factors for good PCR; optimal primers should hybridize only to the target sequence, particularly when complex genomic DNA is used as the template. Amplification problems can arise due to primers annealing to repetitious sequences (retrotransposons, DNA transposons, or tandem repeats). Alternative product amplification can also occur when primers are complementary to inverted repeats and produce multiple bands. This is unlikely when primers have been designed using specific DNA sequences (unique PCR). However, the generation of inverted repeat sequences is exploited in two common generic DNA fingerprinting methods—Random amplified polymorphic DNA (RAPD) and Arbitrarily Primed (AP)-PCR [32, 33]. Because only one primer is used in these PCR reactions, the ends of the products must be reverse complements and thus can form stem-loops.

    The techniques of inter-retrotransposon amplification polymorphism (IRAP), retrotransposon-microsatellite amplification polymorphisms (REMAP), inter-MITE amplification [34, 35], and *Alu*-repeat polymorphism [36, 37] have exploited these highly abundant dispersed repeats as markers. However, primers complementary to repetitious DNA may produce many nonspecific bands in single-primer amplification and compromise the performance of unique PCRs. A homology search of the primer sequence, for example using 'blastn' against all sequences in GenBank or EMBL-Bank, will determine whether the primer is likely to interact with dispersed repeats. Alternatively, one can create a small local specialized library of repeat sequences based on those in Repbase [38] or TREP (http://wheat.pw.usda.gov/ITMI/Repeats/).

The mismatches at the 3′ end of the primers affect target amplification much more than mismatches at the 5′ end. A two base mismatch at the 3′ end of the primer prevents amplification. A single base mismatch as well as several mismatches at the 5′ end of the primer allow amplification, with reduced efficiency of the amplification.

On the other hand, the presence of multiple primer binding sites does not necessarily lead to an alternative amplification, because, for amplification, both primers have to be located close to each other. The close location of the primers at correct orientation to each other and efficient binding of each DNA target determines the probability of alternative amplification.

By default, the FastPCR performs a nonspecific binding test for each given sequence. Additionally, the software allows this test to be performed against a reference sequence or sequences (e.g., BAC, YAC) or one's own database. Primers that bind to more than one location on current sequences will be rejected. Even though the nonspecific primer binding test is performed as a default for all primers, the user may cancel the operation. Identification of secondary binding sites, including mismatched hybridization, is normally performed by considering the similarity of the primer to targets along the entire primer sequence. An implicit assumption is that stable hybridization of a primer with the template is a prerequisite for priming by DNA polymerase. The software pays particular attention to the 3′ end portion of the primer and calculates the similarity of 3′ end of the primer to target (the length is chosen by the user) to determine the stability of the 3′-terminus. The secondary nonspecific primer binding test is based on a quick, non-gapped local alignment (that allows one mismatch within a hash index of 9-mers) screening between the reference and input sequence.

The software offers flexible specificity stringency options. User can specify the number of mismatches that primers must have to unintended targets at 3′ end region where these mismatches must be present. The default specificity settings are that at least one mismatch in the last five bases at the 3′ end of primer.

5   Methods

Once the input files are selected or sequence copy-paste to **General Sequence(s)** text editor, the software provides various execution features. Figure 2 provides a primer design window from user perspective.

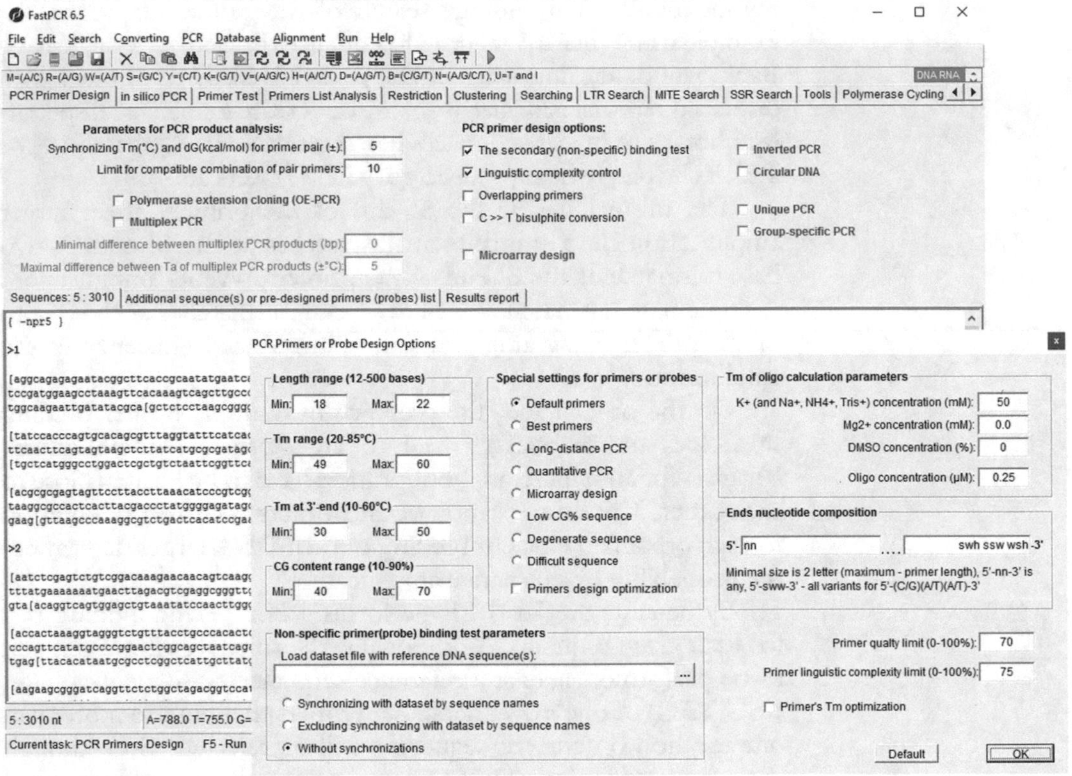

Fig. 2 The "PCR Primers or Probes Design Options" window

5.1 Execution of the Selected Task

User selects the ribbon with the task needed. The program will only perform the selected task. Depending on the task selected, the program will show on the status bar the name of the executive task (Fig. 3) by "**Press F5**." To execute the current task, user can either press the **F5** key or click the arrow on the toolbar with the mouse.

Once the executive task is complete, the result is shown in the **Result report** text editor. Figure 4 shows a sample result visualization window.

5.2 PCR Primer Design Options

The "**PCR Primer Design**" Tab contains various execution options to easily select the type of PCR and most important PCR parameters. Figure 2 shows "**PCR Primers or Probe Design Options**" panel. Once user selects any attribute, the option attribute value field shows the default attributes value, which can then be modified. "**PCR Primers or Probe Design Options**" affects to all sequences. For individual PCR primer design options for each sequence, user can type special commands at the header of sequence (http://primerdigital.com/soft/pcr_help.html). Typically, the user does not need to use commands to manage PCR primer design, all these commands use optionally and only for advanced tasks. User can type at text editor this help command:

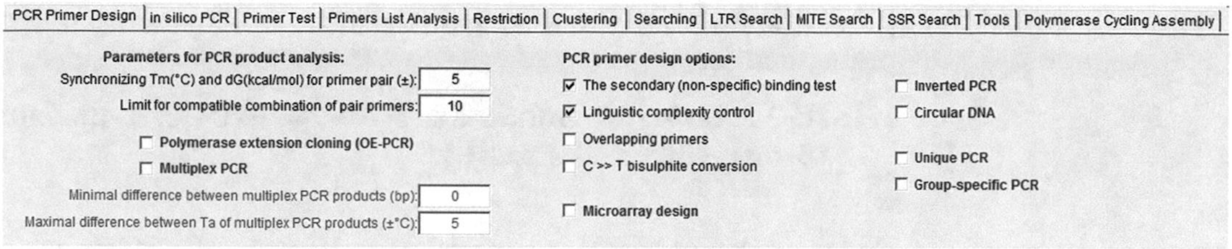

Fig. 3 The ribbon with the tasks. The program will only perform selected task

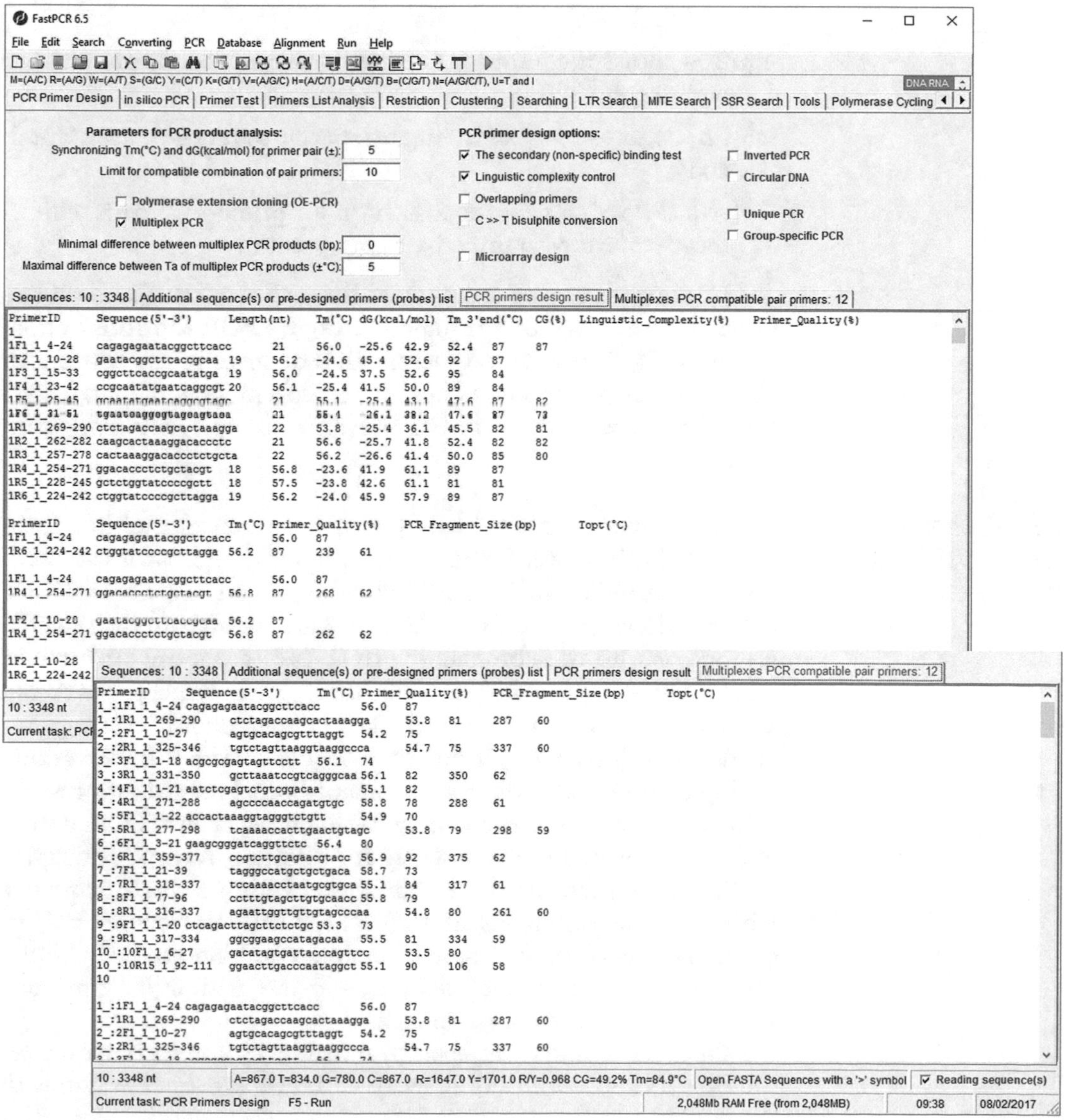

Fig. 4 An example of **Result report** text editor

/? and software replaces it with default global parameters for primer design:

{ -ln20-24 -tm51-60 -3tm30-50 -cg40-70 -q70 -lc75 -npr200 -c5[nn] -c3[swh ssw wsh] },

where

-ln20-24 determines the range of primer length (20–24 bases);

-tm51-60 determines the range of primer Tm (51–60 °C);

-3tm30-50 determines the range of primer Tm at 3′ end (30–50 °C);

-cg40-70 determines the range of primer CG% contents (between 40% and 70%);

-npr200 shows the limit for maximal primers amount designed to each target (200);

-c5[nn] denotes a primer having no specific sequence pattern for 5′ ends;

-c3[swh ssw **wsh**] makes specification for primers' 3′ ends with all these pattern with three bases per pattern.

Commands can be customized for each sequence, and global. Global commands are determined through "**PCR Options**" under menu "**PCR**." In addition, before all sequences in the text-editor (**General Sequence(s)**), as header can specify general parameters for all between {...}, as in the previous example, for help command: /? .

5.3 Examples for Primer Selection Regions

Users can specify, individually for each sequence, multiple locations for both forward and reverse primer design with the commands: -**FpdN1-N2** for forward primers and -**RpdN1-N2** for reverse primers, where from **N1** to **N2** are bases from the selected regions or with the command -**pdN1-N2** (see more at: http://primerdigital.com/soft/pcr_help.html). Alternatively, users can specify the multiple locations for both forward and reverse primers design using [and] inside each sequence: the software allows multiple and independent locations of both forward and reverse primer design inside each of the sequences, whilst PCR design will be performed independently for different targets. Multiplex PCRs can be performed simultaneously within a single sequence with multiple amplicons as well as for different sequences, or combinations of both, i.e., all possible combinations of [and] inside the sequence(s). By default, software is designing primers within the entire sequence length.

Optionally, users can specify, individually for each sequence, multiple locations for both forward and reverse primer design with the commands:

1. The same location for both Forward and Reverse primers will be designed in the central [nnnnnnnnn] part ("[ ]" is Used only once):

       ....[nnnnnnnnnn].......

2. Different locations for Forward and Reverse primers; Forward primers will be chosen inside [**1nnnnnn**] location and Reverse primers inside [**2nnnnnn**] location ("[ ]" is used twice):

   ....[**1nnnnnn**]....[**2nnnnnn**].....

3. Designed primers must flank the central]**nnnnnn**[; Forward primers will be chosen from 1 to " **A**]" bases and Reverse primers will be chosen from "[**C**" base to the end of sequence:

   ......**A**]nnnnnn[**C**.....

4. Design primers with overlapping part [**nnnnnn**] for Forward and Reverse primers; Forward primers will be chosen from [**A to n**] bases and Reverse primers will be chosen from [**n base to C**]:

   ......[**A**.....[**nnnnnn**]......**C**]....

The software allows selecting any amount of independent PCR primers (probe) designing tasks for each sequences using multiple combinations of "[...]" and -**FpdN1-N2**, -**RpdN1-N2** or -**pdN1-N2** commands.

Multiplex PCR can be carried out simultaneously within a single sequence with multiple tasks as well as for different sequences or multiple tasks or both cases together.

All possible combinations of "[ ]" (Forward) with "[ ]" (Reverse) within the sequence(s):

1. [ ]
2.] [
3. [ ] [ ]
4. [[ ]]
5. ([ ] [ ])$_n$ or/and ([[ ]])$_n$.

5.4 User-Defined PCR Product Size

The user can specify the PCR product size in a similar way, with the command: (**N1-N2**); these values can be specified in the form of minimum and maximum value for the product size. For example, the (400–500) line defining the PCR product size ranges from 400 to 500 bases. In case a user wants to specify a fixed product size, the command should be a single number, for example (500). As default, the program allows PCR product sizes ranging between 30 and 10,000 bases.

5.5 PCR Set-Up Examples with Individual Commands

1. Prediction of optimal annealing temperature and PCR fragment(s) length for one or more predesigned primers (with -**npd** command, which prohibits the primer design). For a sequence to which these two primers (5′**ggagagtagcttacctcgct**, 5′**cggtaaggttcttcatgc**) have been designed:

   > −**fpr**[**ggagagtagcttacctcgct cggtaaggttcttcatgc**] -**npd**

2. Design of forward and reverse primers with a difference in T_m of about 10 °C for LATE-PCR:

   > -**Ftm50-55** -**Rtm64-68** -**pTMs10**

3. Design of primers with a specific restriction enzyme site at their 3′ end:

-z3eNameEnzyme, -Fz3eNameEnzyme, -Rz3eNameEnzyme, where **NameEnzyme** is the name of the restriction enzyme: -z3eXceI.

The alternative command (-c3NN) is used for special primer location. For example -c3YCATGR is the same as -z3eXceI, as the result: newly designed primers will locate on restriction site for *Xce I* endonucleases and 3′ end of primers will contain this sequence: **5′-(C/T)CATG(A/G)-3′**.

4. Addition of non-template DNA sequences to primer's ends. Add a sequence to 5′ end with command -**5eNN** or -**3eNN**, where **NN** is some sequence from one to more bases, for example:

-F5eCGACG -R5eTTTTTT, adding sequence **CGACG** to Forward primers and sequence **TTTTTT** to Reverse primers at 5′ ends.

5. LAMP primer design examples:

> -**LAMP** -**LN17-24** -**Tm52-62**

where -**LAMP** assay design (two nested primer pairs, four primers based on six nested targets) without using Loop Primers (http://loopamp.eiken.co.jp/e/lamp/loop.html), with primer lengths from 17 to 14 nt (−**LN17-24**), with the melting temperature from 52 to 62 °C (−**Tm52-62**). By default, the distance between 5′ end of F2 and B2 is considered 120–200 bp and the distance between F2 and F3 as well as B2 and B3 is 0–20 bp. The distance for loop forming regions (5′ of F2 to 3′ of F1, 5′ of B2 to 3′ of B1) is 0–40 bp. For LAMP assay design including Loop primer is necessary to apply the command:

> -**LAMP2** -**LN17-24** -**Tm52-62**,

where -**LAMP2**, will indicate the program for the development of LAMP assay design (two nested primer pairs, six primers based on eight nested targets) using Loop Primers (Fig. 5).

6. Design of tiling arrays: design both overlapping and nonoverlapping PCR primer pairs to generate either distinct or overlapping amplicons. Many unique tiling array platforms using both PCR products and short oligonucleotide probes have been created for a variety of applications including whole-genome arrays and specifically targeted arrays encompassing certain classes of genomic regions. FastPCR can automate the design of PCR primers for tiling arrays using both PCR products and short oligonucleotides, and avoiding repetitive regions. The user can specify the lengths of overlapping PCR primer pairs with the command: -**TILLING[N1N2]**; these values can be specified in the form of minimum and maximum value for the overlapping lengths. For example, the target DNA sequence using the **"Overlapping Amplicon"** option:

Fig. 5 An example of LAMP assay design using Loop Primers result

> **(110–250) -TILLING[-100-10]**,

where both **N1** and **N2** negative values allow overlapping between amplicons from 10 to 100 bases, command **(110–250)** defines the amplicons' size range from 110 to 250 bases.

The other option available for PCR primer pairs design is the **"Distinct Amplicon"** option. The amplicons generated could be as close as one base pair apart:

> **(100–250) -TILLING[0100]**,

where both **N1** and **N2** positive values do not allow overlapping amplicons and distance between amplicons ranges from 0 to 100 bases.

5.6 Bisulfite Modified DNA

The **"C >>T bisulphite conversion"** option allows the design of specific PCR primers for in silico bisulfite conversion for both strands—only cytosines not followed by guanidine (CpG methylation) will be replaced by thymines:

```
5'aaCGaagtCC-3'                    5'aaCGaagtTT-3'
| | | | | | | | |          ->       | | | | | |  |
3'ttGCttCagg-5'                    3'ttGCttTagg
```

**5.7 PCR
Primer Design**

The PCR primer design algorithm generates a set of primers with a high likelihood of success in any amplification protocol. All PCR primers designed by FastPCR can be used for PCR, sequencing experiments or isothermal amplification.

The program is able to generate either long oligonucleotides or PCR primers for the amplification of gene-specific DNA fragments of user-defined length. FastPCR provides a flexible approach to designing primers for many applications and for linear and circular sequences. It will check if either primers or probes have secondary binding sites in the input sequences that may give rise to an additional PCR product. The selection of the optimal target region for the design of long oligonucleotides is performed in the same way as for PCR primers. The basic parameters in primer design are also used as a measure of the oligonucleotide quality and the thermodynamic stability of the 3' and 5' terminal bases are evaluated.

The proposal of primer pairs and the selection of the best pairs are both possible. The user can vary the product size or design primer pairs for the whole sequence without specifying parameters by using default or predesigned parameters. The predesigned parameters are specified for different situations: for example, for sequences with low CG content or long distance PCR, or degenerate sequences, or for manual input. A list of best primer candidates and all compatible primer pairs that are optimal for PCR is generated. Users can specify, individually for each sequence, multiple locations for both forward and reverse primer design inside each sequence, whilst PCR design will be performed independently for different targets. Multiplex PCRs can be performed simultaneously within a single sequence with multiple amplicons as well as for different sequences, or combinations of both (Fig. 1).

The program generates primer pairs (and probes) from the input sequences and shows the optimal annealing temperature for each primer pair and the sizes of PCR products, together with information for each designed primer. Results are generated by the program showing the suggested primers and primer pairs in tabulated format for Excel or Open Office. The spreadsheets show the following properties: automatically generated primer name, primer sequence, sequence location, direction, length, melting temperature, CG content (%), molecular weight, molar extinction coefficient, linguistic complexity (%), and PQ. For compatible primer pairs, the annealing temperature and PCR product size are also provided.

5.8 Multiplex PCR Primer Design

Multiplex PCR is an approach commonly used to amplify several DNA target regions in a single reaction. The simultaneous amplification of many targets reduces the number of reactions that needs to be performed; multiplex PCR thus increases throughput efficiency. The design of multiplex PCR assays can be difficult because it involves extensive computational analyses of primer pairs for cross interactions. The multiplex PCR algorithm is based on the fast non-recursion method, with the software performing checks on product size compatibility (if necessary), the melting and annealing (T_a) temperatures, dG compatibility and cross-dimer interaction for all primers. To achieve uniform amplification of the targets, the primers must be designed to bind with equal efficiencies to their targets. FastPCR can quickly design a set of multiplex PCR primers for all the input sequences and/or multiplex targets within each sequence. PCR conditions may need to be adjusted; for example, the annealing temperature increased or lowered so that all products are amplified with equal efficiency. To achieve this, most existing multiplex primer design packages use primer melting temperature. In practical terms, the design of nearly identical T_a's and T_m's is important. The melting temperatures of the PCR products are also important, since these are related to annealing temperature values. The T_m of a PCR product directly depends on its GC content and length; short products are more efficiently amplified at low PCR annealing temperatures (100 bp, 50–55 °C) than long products (>3000 bp, 65–72 °C). For most multiplex PCRs, there is usually a small variation (up 5 °C) between the optimal T_a's of all primer pairs and PCR products. The annealing temperature must be optimal in order to maximize the likelihood of amplifying the target genomic sequences while minimizing the risk of nonspecific amplification. Further improvements can be achieved by selecting the optimal set of primers that maximize the range of common T_m's. Once prompted, FastPCR calculates multiplex PCR primer pairs for given target sequences. The speed of calculation depends on the number of target sequences and primer pairs involved.

An alternative way to design compatible multiplex PCR primer pairs is to use predesigned primers as references for the design of new primers. The user can also select input options for the PCR products such as the minimum product size difference between the amplicons. One can set primer design conditions either individually for each given sequence or use common values. The individual setting has a higher priority for PCR primer or probe design than do the general settings. The results include primers for individual sequences, primers compatible together, the product sizes, and annealing temperatures. Because clear differentiation of the products is dependent on using compatible primer pairs in the single reactions, the program recovers all potential variants of primer combinations for analyses of the chosen DNA regions and provides, in tabular form, their compatibility with information including primer-dimers, cross-hybridization, product size overlaps, and similar alternative primer pairs based on T_m. The user may choose those alternative compatible primer pair combinations that provide the desired product sizes. Using the program, researchers can select predesigned primer pairs from a target for their desired types of PCR reactions by changing the filtering conditions as mentioned above. For example, a conventional multiplex PCR for gel-electrophoresis analysis requires differently sized (at least by 10 bp) amplicons for a set of target genes, so the value for the minimum size difference between PCR products can be selected.

In addition to the need to avoid same-sized amplicons, multiplex PCR must also minimize the generation of primer dimers and secondary products, which becomes more difficult with increasing numbers of primers in a reaction. To avoid the problem of nonspecific amplification, FastPCR allows the selection of primer pairs that give the most likelihood of producing only the amplicons of the target sequences by choosing sequences which avoid repeats or other motifs. The program also allows the user to design not only compatible pairs of primers, but also compatible single primers for different targets or sequences.

The input sequence can be made of either a single with minimum two internal tasks or many sequences with or without internal tasks. Most of the parameters on the interface are self-explanatory. Optionally, the user is asked to provide the sequence and select oligonucleotides designing parameters.

On the **PCR Primer Design** tab the user chooses the **Multiplex PCR** option and selects the limit for multiple PCR compatible combination of pair primers (default 100 primer pairs), minimal difference between multiplex PCR products (default 10 bp) and maximal difference between T_a/dG of multiplex PCR products (default ±5 °C). After specifying inputs and PCR primers design options, the user can execute the PCR primers design task. Once the primer set's design is complete, the result will appear in two **Result** text editors: **PCR primer design result** and **Multiplex PCR compatible pair primers**. Figure 4 shows the access to PCR primer design output. The result text editors **PCR primer design result** displays the individual PCR primers design data, including the primers list and the compatible primer pairs for all the sequences and their internal tasks where suitable primers are found. Second **Multiplex PCR compatible pair primers** text editors collects final search results that are presented as a list of the sets of the compatible primer pairs for multiplex PCR.

5.9 Group-Specific PCR Primers

Group-specific amplification, also called family-specific and sequence-specific amplification, is an important tool for comparative studies of related genes, sequences, and genomes that can be applied to studies of evolution, especially for gene families and for cloning new related sequences. Specific targets such as homological genes or transposable elements can be amplified to uncover DNA polymorphisms associated with these sequences or other genetic investigations. The overall strategy of designing group-specific PCR primers uses a hash index of 12-mers to identify common regions in target sequences, following standard PCR design for the current sequence, and then testing complementarity of these primers to the other sequences. FastPCR performs either multiple sequence alignment or accepts alignment sequence input, giving it the flexibility to use a different strategy for primer design. Also, it can design both degenerate and nondegenerate PCR primers to amplify a conserved or polymorphic region of all related sequences.

The software designs large sets of universal primer pairs for each given sequence, identifies conserved regions, and generates suitable primers for all given targets. The steps of the algorithm are performed automatically and the user can influence the general options for primer design options. The FastPCR will work with any source of sequence as long as it is possible to find short (minimum 12 nt) consensus sequences among the sets. The quality of primer design is dependent on sequence relationships, phylogenetic similarity, and suitability of the consensus sequence for the design of good primers. The software is able to generate group-specific primers for each set of sequences independently, which are suitable for all sequences. Primer alignment parameters for group-specific PCR primers are similar to those used for in silico PCR.

On the **PCR Primer Design** tab the user chooses the **Group-specific PCR** option. After specifying inputs and PCR primers design options, the user can execute the PCR primers design task.

The program takes either multiple separate DNA sequences in either FASTA or alignment formats.

Once the primer set design is complete, the result will appear in the **Result** text editor: **PCR primer design result**. The result text editor **PCR primer design result** displays the individual group-specific PCR primer design data, including the primer list and compatible primer pairs for all the sequences and their internal tasks where suitable primers are found. In the case an alignment has been input, the result text editors display only the group-specific PCR primer design set, including both degenerate and nondegenerate primers in the primer list, as well as compatible primer pairs for all the sequences.

5.10 Simple Sequence Repeat (SSR) Locus Search and PCR Primer Design

Simple sequence repeats (SSRs, or microsatellites) are short tandem repeats of one or more bases. Microsatellites are ubiquitously distributed throughout eukaryotic genomes, often highly polymorphic in length, and thereby an important class of markers for population genetic studies. Our approach to SSR searching is to analyse low complexity regions by using linguistic sequence complexity. This method allows the detection of perfect and imperfect SSRs with a single, up to 10-base, repeat motif. Each entry sequence is processed for identification of SSRs and the SSR flanks are used to design compatible forward and reverse primers for their amplification by PCR.

FastPCR identifies all SSRs within each entry sequence and designs compatible PCR primer pairs for each SSR locus. The default PCR primer design parameters are that the primers must be within 100 bases from either side of the identified SSR. Often the sequences available around SSR loci are not suitable for designing good primers and the user can increase or decrease the distance from either side to find more efficient and compatible primer pairs. The capabilities of FastPCR make it a complete bioinformatics tool for the use of microsatellites as markers, from discovery through to primer design. For example, the user can specify PCR primer design to SSR loci within 200 bp around SSR, with command: -ssr/200. The software finds all SSR sites and then will design PCR primers and compatible primer pairs independently for each SSR loci.

5.11 Polymerase Cycling Assembly

The application to make long synthetic DNA molecules rely on the in vitro assembly of a set of short oligonucleotides, either by ligase chain reaction (LCR) [39] or by assembly PCR [40]. These oligonucleotides should be adjacent on the same strand and overlap the complementary oligonucleotides from the second strand. There are several major parameters to designing oligonucleotides for gene synthesis by LCR or assembly PCR: first, the oligonucleotides should share about similar T_m value; second, a given oligonucleotides sequence should be unique to avoid multiple nonspecific binding that conduct to incorrect assembly. The software must dynamically choose the length of the oligonucleotides to ensure both the specificity and the uniform T_m. Our algorithm is able to design oligonucleotides for long sequences containing repeats and to minimize their potential nonspecific hybridization during 3′ end extension in PCR. For long sequence assembly, oligonucleotide design starts from the 5′ end of a given sequence; the oligonucleotide length is dynamically changed until a unique 3′ end has been found and T_m of oligonucleotide has reached the T_m threshold. All oligonucleotides are designed without gaps between them. The other strand is used for design of the overlapping oligonucleotides with the same algorithm as above but with the T_m of the overlapping regions reaching the $T_m - 15$ °C threshold. The composition of the sequence at the 3′ terminus is important because stability at the 3′ terminus in the oligonucleotide complexes will improve the specificity of extension by the polymerase. To reduce nonspecific polymerase extension and ligation, the algorithm chooses only unique sequences for the 3′ terminus. Minimally, the last two nucleotides at the 3′ terminus must not be complementary to the nonspecific target. Other complementary regions, apart from the 3′ terminus, are not important for assembling multiple fragments by PCR and ligation.

The input sequence can comprise either a single or many sequences. Most of the parameters on the interface are self-explanatory. The user is asked to provide the sequence and select oligonucleotide designing parameters. On the **Polymerase Cycling Assembly** tab, in the **Polymerase cycling assembly options**, the user can choose minimal (40–100 nt) and maximal (80–200 nt) oligonucleotide lengths, and minimal overlapping lengths (20–40 nt); by default the oligonucleotides' length ranges from 50 to 90 nt and 20 nt for overlapping length. The interface allows changing T_m calculation parameters (salt, Mg^{2+} and oligonucleotide concentration).

The searching process runs after pressing **F5** or from menu bar or from toolbox. The research result is presented as a list of oligonucleotides for both chains. On both strands, all oligonucleotides are overlapping with gap between neighbor oligonucleotides. Certain oligonucleotides will overlap two oligonucleotides from the complementary strand. The algorithm pays attention to avoid non-specific oligonucleotide hybridization to repeated regions. When it is not possible to design primers outside of repeated sequences, it will be difficult to find short specific oligonucleotides. The solution to this problem is to divide the sequence into short segments, design a set of oligonucleotides for each segment independently, and then combine all these segments in the second PCR for final amplification. Figure 6 shows a sample result visualization window.

5.12 Polymerase Extension PCR for Fragment Assembly

Sequence-independent cloning, including ligation-independent cloning requires generation of complementary single-stranded overhangs in both the vector and insertion fragments. Similarly, multiple fragments can be joined or concatenated in an ordered manner using overlapping primers in PCR. Annealing of the complementary regions between different targets in the primer overlaps allows the polymerase to synthesize a contiguous fragment containing the target sequences during thermal cycling, a process called "overlap extension PCR" (OE-PCR) [41]. The efficiency depends on the T_m and on the length and uniqueness of the overlap. To achieve this, FastPCR designs compatible forward and reverse primers at the ends of each fragment, and then extends the 5′ end of primers using sequences from the primers of the fragment that will be adjacent in the final product.

The input sequence can be made of either a single or many sequences. The user needs to take special attention to the preparation of sequences for assembly.

The users can specify the locations for both forward and reverse primers design using "[ ]" to define each sequence ends. The defined regions will be used by the program for designing the overlapping primers.

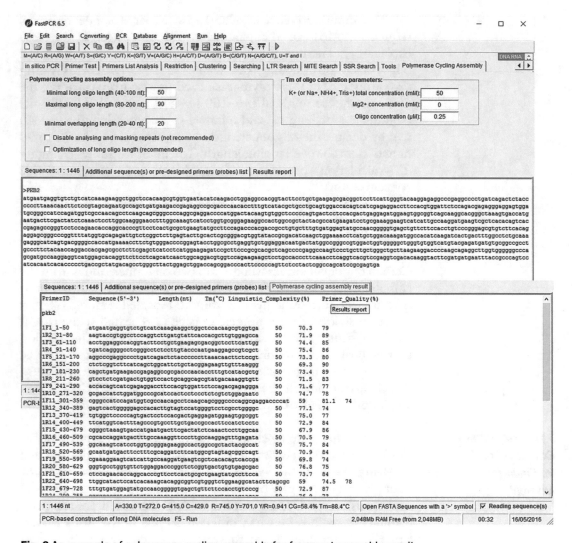

Fig. 6 An example of polymerase cycling assembly for fragment assembly result

The program selects the overlapping area so that the primers from overlapping fragments are similar in size and in their optimal annealing temperature. The program adds the required bases so that the T_m of the overlap is similar to or higher than the T_m of the initial primers. Primers are tested for dimers within the appropriate primer pair.

On the **PCR Primer Design** tab the user chooses the **Polymerase extension cloning (OE-PCR)** option and the limit for multiple PCR compatible combination of pair primers (default is 100). After specifying inputs and PCR primers design options, the user can execute the searching task. Once the primer set's design is complete, the result will appear in two **Result** text editors: **PCR primer design result** and **PCR fragments assembling compatible pair primers**.

The result text editors **PCR primer design result** displays the individual PCR primer design data, including the primer list and compatible primer pairs for all sequences where suitable primers are found. The **PCR fragments assembling compatible pair primers** text editor collects the final search result, which is presented as a list of sets of the compatible primer pairs for individual fragments amplification and assembly.

5.13 Primer Analyses

Individual and sets of primers are evaluated using software. They calculate primer T_m's using default or other formulae for normal and degenerate nucleotide combinations, CG content, extinction coefficient, unit conversion (nmol per OD), mass (µg per OD), molecular weight, linguistic complexity, and consider primer PCR efficiency. Users can select either DNA or RNA primers (online Java: PrimerAnalyser) with normal or degenerate oligonucleotides or modifications with different labels (for example inosine, uridine, or fluorescent dyes). Tools allow the choice of other nearest-neighbor thermodynamic parameters or nonthermodynamic T_m calculation formulae.

For LNA modifications the four symbols: dA = E, dC = F, dG = J, dT = L are used. Both programs perform analyses on-type, which allow users to see the results immediately on screen. They can also calculate the volume of solvent required to attain a specific concentration from the known mass (mg), OD, or moles of dry oligonucleotide.

All primers are analyzed for intra- and inter-primer interactions to form dimers. Primer(s) can efficiently hybridize using the 5′ end or middle of the sequences. Even though such interactions are not efficiently extended by DNA polymerase, their formation reduces the effective primer concentration available for binding to the target and their presence can strongly inhibit PCR, since double-stranded DNA at high concentrations is a strong inhibitor of DNA polymerase (Fig. 7).

Example for a primer complementarity test including mismatches: when it is necessary to determine the annealing stability at the binding site and the T_m for primers with mismatches, the user may use the in silico PCR task. Another way to solve this problem is by analyzing the two primers in the task **Primer Test**, where one primer is analyzed, and the second complementary strand of it. To do so, it is necessary to convert the primer sequence into complementary and inverse chain and exclude the mismatches. The program will identify self-dimer and calculate the T_m including mismatches:

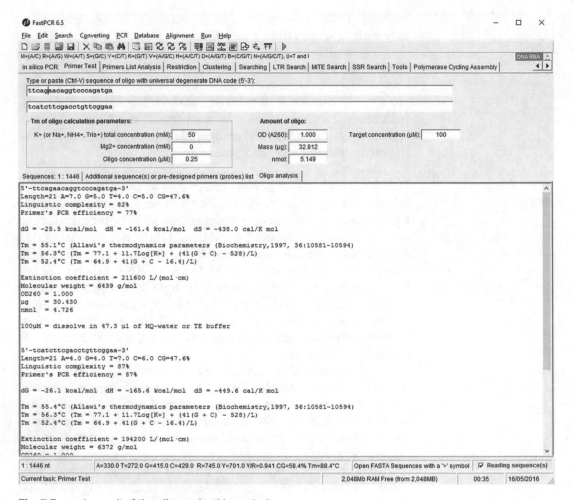

Fig. 7 Example result of the oligonucleotide analysis

```
Tm = 38.8 °C
5'-ttcagaacaggtcccagatga-3'
Length=21 A=7.0 G=5.0 T=4.0 C=5.0 CG=47.6%
Linguistic complexity = 82%
Primer's PCR efficiency = 77%
Tm = 55.1 °C

5'-tcatcttcgacctgttcggaa-3'
Length=21 A=4.0 G=4.0 T=7.0 C=6.0 CG=47.6%
Linguistic complexity = 87%
Primer's PCR efficiency = 87%
Tm = 55.4 °C

Dimers found between primers:

<-aaggcttgtccagcttctact-5
  |||:||||||||||  ||||||
```

```
5-ttcagaacaggtcccagatga->
Tm=38.8 °C
```

Acknowledgments

Web tools are available free, provided for noncommercial research and education use only. They may not be reproduced or distributed for commercial use. This work was supported by the companies PrimerDigital Ltd.

References

1. Walker-Daniels J (2012) Current PCR methods. Mater Methods 2:119. doi:10.13070/mm.en.2.119

2. Tisi LC, Gandelman O, Kiddle G, Mcelgunn C (2010) Nucleic acid amplification. Canada Patent CA2417798

3. Notomi T, Okayama H, Masubuchi H, Yonekawa T, Watanabe K, Amino N, Hase T (2000) Loop-mediated isothermal amplification of DNA. Nucleic Acids Res 28(12):e63. doi:10.1093/nar/28.12.e63

4. Rozen S, Skaletsky H (2000) Primer3 on the WWW for general users and for biologist programmers. Methods Mol Biol 132:365–386

5. Kalendar R, Lee D, Schulman AH (2014) FastPCR software for PCR, in silico PCR, and oligonucleotide assembly and analysis. In: Valla S, Lale R (eds) DNA cloning and assembly methods, Methods in molecular biology, vol 1116. Humana, New York, pp 271–302. doi:10.1007/978-1-62703-764-8_18

6. Kalendar R, Lee D, Schulman AH (2011) Java web tools for PCR, in silico PCR, and oligonucleotide assembly and analysis. Genomics 98(2):137–144. doi:10.1016/j.ygeno.2011.04.009

7. Marshall OJ (2004) PerlPrimer: cross-platform, graphical primer design for standard, bisulphite and real-time PCR. Bioinformatics 20(15):2471–2472. doi:10.1093/bioinformatics/bth254

8. Owczarzy R, Tataurov AV, Wu Y, Manthey JA, McQuisten KA, Almabrazi HG, Pedersen KF, Lin Y, Garretson J, McEntaggart NO, Sailor CA, Dawson RB, Peek AS (2008) IDT SciTools: a suite for analysis and design of nucleic acid oligomers. Nucleic Acids Res 36:163–169. doi:10.1093/nar/gkn198

9. Bekaert M, Teeling EC (2008) UniPrime: a workflow-based platform for improved universal primer design. Nucleic Acids Res 36(10):e56. doi:10.1093/nar/gkn191

10. Ye J, Coulouris G, Zaretskaya I, Cutcutache I, Rozen S, Madden TL (2012) Primer-BLAST: a tool to design target-specific primers for polymerase chain reaction. BMC Bioinformatics 13:134. doi:10.1186/1471-2105-13-134

11. Smykal P, Kalendar R, Ford R, Macas J, Griga M (2009) Evolutionary conserved lineage of Angela-family retrotransposons as a genome-wide microsatellite repeat dispersal agent. Heredity (Edinb) 103(2):157–167. doi:10.1038/hdy.2009.45

12. Giegerich R, Meyer F, Schleiermacher C (1996) GeneFisher--software support for the detection of postulated genes. Proc Int Conf Intell Syst Mol Biol 4:68–77

13. Gadberry MD, Malcomber ST, Doust AN, Kellogg EA (2005) Primaclade--a flexible tool to find conserved PCR primers across multiple species. Bioinformatics 21(7):1263–1264. doi:10.1093/bioinformatics/bti134

14. Nomenclature Committee of the International Union of Biochemistry (NC-IUB) (1984) Nomenclature for incompletely specified bases in nucleic acid sequences. http://www.chem.qmul.ac.uk/iubmb/misc/naseq.html

15. Allawi HT, SantaLucia J Jr (1997) Thermodynamics and NMR of internal G.T mismatches in DNA. Biochemistry 36(34):10581–10594. doi:10.1021/bi962590c

16. SantaLucia J (1998) A unified view of polymer, dumbbell, and oligonucleotide DNA nearest-neighbor thermodynamics. Proc Natl Acad Sci U S A 95(4):1460–1465

17. Le Novere N (2001) MELTING, computing the melting temperature of nucleic acid duplex. Bioinformatics 17(12):1226–1227. doi:10.1093/bioinformatics/17.12.1226

18. Bolton ET, McCarthy BJ (1962) A general method for the isolation of RNA complementary to DNA. Proc Natl Acad Sci U S A 48(8):1390–1397

19. Guedin A, Gros J, Alberti P, Mergny JL (2010) How long is too long? Effects of loop size on G-quadruplex stability. Nucleic Acids Res 38(21):7858–7868. doi:10.1093/nar/gkq639

20. Wallace RB, Shaffer J, Murphy RF, Bonner J, Hirose T, Itakura K (1979) Hybridization

of synthetic oligodeoxyribonucleotides to ΦX 174 DNA: the effect of single base pair mismatch. Nucleic Acids Res 6(11): 3543–3558. doi:10.1093/nar/6.11.3543

21. von Ahsen N, Wittwer CT, Schutz E (2001) Oligonucleotide melting temperatures under PCR conditions: nearest-neighbor corrections for Mg2+, deoxynucleotide triphosphate, and dimethyl sulfoxide concentrations with comparison to alternative empirical formulas. Clin Chem 47(11):1956–1961

22. Gabrielian A, Bolshoy A (1999) Sequence complexity and DNA curvature. Comput Chem 23(3–4):263–274. doi:10.1016/S0097-8485(99)00007-8

23. Orlov YL, Potapov VN (2004) Complexity: an internet resource for analysis of DNA sequence complexity. Nucleic Acids Res 32(Web Server issue):628–633. doi:10.1093/nar/gkh466

24. Gilson MK, Given JA, Bush BL, McCammon JA (1997) The statistical-thermodynamic basis for computation of binding affinities: a critical review. Biophys J 72(3):1047–1069. doi:10.1016/S0006-3495(97)78756-3

25. Peyret N, Seneviratne PA, Allawi HT, SantaLucia J Jr (1999) Nearest-neighbor thermodynamics and NMR of DNA sequences with internal a.A, C.C, G.G, and T.T mismatches. Biochemistry 38(12):3468–3477. doi:10.1021/bi9825091

26. Watkins NE Jr, SantaLucia J Jr (2005) Nearest-neighbor thermodynamics of deoxyinosine pairs in DNA duplexes. Nucleic Acids Res 33(19):6258–6267. doi:10.1093/nar/gki918

27. Sen D, Gilbert W (1992) Guanine quartet structures. Methods Enzymol 211:191–199

28. Il'icheva IA, Florent'ev VL (1992) Four-stranded complexes of oligonucleotides--quadruplexes. Mol Biol 26(3):512–531

29. Shing Ho P (1994) The non-B-DNA structure of d(CA/TG)n does not differ from that of Z-DNA. Proc Natl Acad Sci U S A 91(20):9549–9553

30. Kypr J, Kejnovska I, Renciuk D, Vorlickova M (2009) Circular dichroism and conformational polymorphism of DNA. Nucleic Acids Res 37(6):1713–1725. doi:10.1093/nar/gkp026

31. SantaLucia J Jr, Hicks D (2004) The thermodynamics of DNA structural motifs. Annu Rev Biophys Biomol Struct 33:415–440. doi:10.1146/annurev.biophys.32.110601.141800

32. Williams JGK, Kubelik AR, Livak KJ, Rafalski JA, Tingey SV (1990) DNA polymorphisms amplified by arbitrary primers are useful as genetic-markers. Nucleic Acids Res 18(22):6531–6535. doi:10.1093/nar/18.22.6531

33. Welsh J, Mcclelland M (1990) Fingerprinting genomes using pcr with arbitrary primers. Nucleic Acids Res 18(24):7213–7218. doi:10.1093/nar/18.24.7213

34. Kalendar R, Schulman A (2006) IRAP and REMAP for retrotransposon-based genotyping and fingerprinting. Nat Protoc 1(5):2478–2484. doi:10.1038/nprot.2006.377

35. Chang RY, O'Donoughue LS, Bureau TE (2001) Inter-MITE polymorphisms (IMP): a high throughput transposon-based genome mapping and fingerprinting approach. Theor Appl Genet 102(5):773–781. doi:10.1007/s001220051709

36. Nelson DL, Ledbetter SA, Corbo L, Victoria MF, Ramirez-Solis R, Webster TD, Ledbetter DH, Caskey CT (1989) Alu polymerase chain reaction: a method for rapid isolation of human-specific sequences from complex DNA sources. Proc Natl Acad Sci U S A 86(17):6686–6690

37. Sinnett D, Deragon JM, Simard LR, Labuda D (1990) Alumorphs—human DNA polymorphisms detected by polymerase chain reaction using Alu-specific primers. Genomics 7(3):331–334

38. Jurka J, Kapitonov VV, Pavlicek A, Klonowski P, Kohany O, Walichiewicz J (2005) Repbase update, a database of eukaryotic repetitive elements. Cytogenet Genome Res 110(1–4):462–467. doi:10.1159/000084979

39. Heckman KL, Pease LR (2007) Gene splicing and mutagenesis by PCR-driven overlap extension. Nat Protoc 2(4):924–932. doi:10.1038/nprot.2007.132

40. Higasa K, Hayashi K (2002) Ordered catenation of sequence-tagged sites and multiplexed SNP genotyping by sequencing. Nucleic Acids Res 30(3):E11

41. Quan J, Tian J (2009) Circular polymerase extension cloning of complex gene libraries and pathways. PLoS One 4(7):e6441. doi:10.1371/journal.pone.0006441

Chapter 3

Long Fragment Polymerase Chain Reaction

Eng Wee Chua, Simran Maggo, and Martin A. Kennedy

Abstract

Polymerase chain reaction (PCR) is an oft-used preparatory technique in amplifying specific DNA regions for downstream analysis. The size of an amplicon was initially limited by errors in nucleotide polymerization and template deterioration during thermal cycling. A variant of PCR, designated *long-range PCR*, was devised to counter these drawbacks and enable the amplification of large fragments exceeding a few kb. In this chapter we describe a protocol for long-range PCR, which we have adopted to obtain products of 6.6, 7.2, 13, and 20 kb from human genomic DNA samples.

Key words Long-range polymerase chain reaction (PCR), Agarose gel electrophoresis, *Taq* DNA polymerase, Proofreading enzyme, Thermal cycling, Long amplicons

1 Introduction

Polymerase chain reaction (PCR) is an indispensable tool in molecular biology; without it much of the contemporary work entailing DNA analysis would not have been feasible. The principle of the technique is rather simple: DNA samples are subjected to alternating steps of heating and cooling that cyclically initiate and terminate DNA synthesis. Complementary DNA strands are first separated by heat to allow the subsequent *annealing* of a pair of short oligonucleotides, designated *primers*. A thermostable DNA polymerase adds nucleotides to the *primed* DNA templates, creating multiple copies of the demarcated target region. The resultant products or *amplicons* are then bound to a nucleic acid stain, which fluoresces upon excitation by ultraviolet light, and visualized by agarose gel electrophoresis [1, 2].

The synthesis of long products exceeding a few kb was initially hindered by several drawbacks of the technique that were exacerbated by the extended target length. Two major alterations to the PCR protocol were made to counter these limitations. A proofreading enzyme was added to correct nucleotide misincorporation that would otherwise thwart the elongation of nascent DNA

Lucília Domingues (ed.), *PCR: Methods and Protocols*, Methods in Molecular Biology, vol. 1620,
DOI 10.1007/978-1-4939-7060-5_3, © Springer Science+Business Media LLC 2017

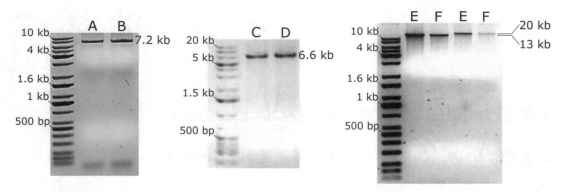

Fig. 1 Amplification of long PCR fragments from three important pharmacogenes: *CYP2A6* (*left panel*), *CYP2D6* (*middle panel*), and *CYP2C19* (*right panel*). *a–f* represent different human genomic DNA samples

strands and therefore constrain the final product size. An alkaline buffer system was adopted to curb heat-induced depurination and preserve the integrity of long amplicons [3–5]. By adopting the protocol detailed in this chapter, we have obtained PCR products of 6.6, 7.2, 13, and 20 kb from human genomic DNA samples (Fig. 1).

2 Materials

2.1 PCR

1. Primer stocks reconstituted in 1× Tris–EDTA to a concentration of 50–100 μM, stored at −20°C (*see* **Note 1**). These should be diluted to 5–10 μM, for use as working solutions, and kept at 4 °C (*see* **Note 2**).

2. Long-range enzyme mix consisting of a *Taq* DNA polymerase and a proofreading enzyme (*see* **Note 3**).

3. Manufacturer-supplied 10× reaction buffer (*see* **Note 4**).

4. Magnesium chloride solution (25–50 mM; *see* **Note 4**).

5. Deoxynucleoside triphosphate (dNTP) mix (8 mM; *see* **Note 4**).

6. Genomic DNA or other types of templates such as plasmid DNA (*see* **Note 5**).

7. Additives such as betaine or dimethylsulfoxide (DMSO) (*see* **Note 6**).

8. Horizontal laminar flow hood.

9. Thermal cycler.

2.2 Agarose Gel Electrophoresis

1. Low electroendosmosis agarose gel powder.

2. 10× Tris–acetic acid–EDTA (TAE) buffer (as running buffer and for dissolving gel): In a 1 L bottle, add 48.4 g of Tris base, 3.7 g of ethylenediaminetetraacetic acid (EDTA) disodium salt

or 20 mL of 0.5 M EDTA solution, 11.4 mL of glacial acetic acid (*see* **Note 7**), and sterile water up to 1 L. Replace the lid and shake the bottle to evenly mix the ingredients. There is no need to adjust the pH of the buffer; the EDTA powder should dissolve readily.

3. Agarose gel electrophoresis apparatus.

4. Ultraviolet transilluminator with a gel documentation unit.

3   Methods

3.1   Primer Design

The presence of pseudogenes or genes with similar consensus sequence is a common problem when trying to identify specific single-nucleotide polymorphisms (SNPs). One way to precisely target SNPs in a particular gene is to ensure the primers will amplify the target gene only. For instance, the *CYP2A6* and *CYP2A7* genes are 96.5% identical with respect to their coding nucleotide sequences, and identifying SNPs specific to *CYP2A6* using small fragment PCR can often result in the amplification of regions of the *CYP2A7* gene [6]. The design of primers for the specific amplification of the *CYP2A6* gene is described below as an example.

3.1.1   Obtaining the Reference Sequence

1. Locate the gene, *CYP2A6* (Gene ID 1548), which in this instance has been reviewed and collated into the RefSeq database: http://www.ncbi.nlm.nih.gov/gene/ [7].

2. Select the appropriate genome assembly, in this case we have chosen the GRCh37.p13 primary assembly, and obtain the FASTA sequence for *CYP2A6*. The gene length is approximately 6.9 kb.

3. Update the FASTA sequence shown to include 500 bp on both 3′ and 5′ ends. This can be done by manipulating the chromosome position numbers. The total length of the updated *CYP2A6* FASTA sequence is now approximately 7.9 kb.

3.1.2   Obtaining Specific Primer Sequences (See Fig. 2)

1. Copy or upload the updated FASTA sequence onto an online tool for primer design, Primer-BLAST; this tool can be accessed at http://www.ncbi.nlm.nih.gov/tools/primer-blast/. The operation of the primer design feature is driven by Primer 3 [8, 9].

2. Specify the desired primer positions relative to the inserted FASTA sequence. As we have allowed for a 500-bp window on either side of the gene, forward primers can start anywhere from 0 to 300 and reverse primers can start anywhere between 7500 and 7800. Also, the primers are positioned in such a manner that they are 150 bp away from the intron–exon boundaries and that the resultant product would cover the entire *CYP2A6* gene.

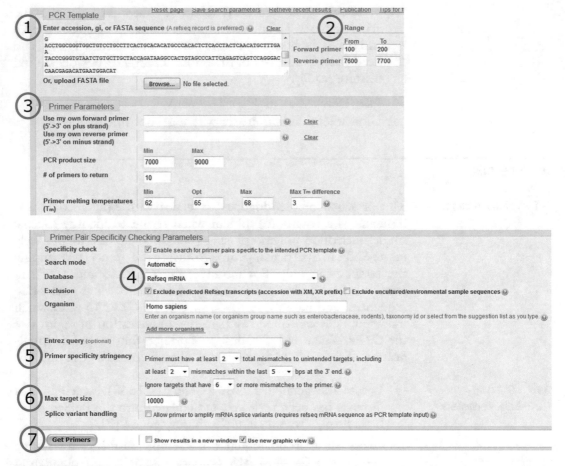

Fig. 2 A simple guide to primer design using Primer-BLAST, a free Web-based tool. *Circled numbers* refer to the steps described in the text

3. Adjust the primer parameters to suit the requirements for long-range PCR (*see* **Note 8**). Primers for long-range PCR should be designed to have a melting temperature of at least 65 °C so that a two-step cycling protocol can be employed, whereby both annealing and extension temperatures are set to 68 °C.

4. Select *Refseq mRNA* as the reference sequence database and exclude predicted Refseq transcripts. The target organism should be *Homo sapiens* in this case.

5. The parameters for primer specificity stringency can be left unchanged. All primers are typically designed to contain at least two mismatches to nontarget templates at the 3′ end. Increasing the minimum number of mismatches to unwanted targets augments primer specificity but also diminishes the likelihood of the tool finding suitable primer pairs.

6. Adjust the maximum target size so that it equals or exceeds the estimated product size.

7. Click on the *Get Primers* button to initiate the tool.

3.2 PCR

1. Prepare all reaction mixes inside a horizontal laminar flow hood. Also, it is mandatory to have strictly separate areas for pre- and post-PCR work to reduce the chances of contamination.

2. Irradiate the equipment and PCR reagents used for reaction setup with ultraviolet light for about 15–20 min, in order to minimize the risk of carryover contamination; this may include the pipettes, pipette tips, microfuge tubes, and PCR strip tubes [10, 11].

3. While waiting for the decontamination procedure to finish, set up the PCR protocol in a thermal cycler (*see* **Notes 9** and **10**).

4. Add, in a 0.7 mL microfuge tube (or larger if required), the following reagents: reaction buffer, magnesium chloride, dNTP mix, primers, betaine or DMSO (if necessary), and enzyme mix (*see* **Note 11**).

5. Briefly flick the tube to mix all reaction components and spin down the mixture with a 5- to 10-s pulse in a microcentrifuge.

6. Divide the master mix into equal-volume aliquots and transfer these into individual PCR strip tubes (*see* **Notes 12** and **13**). Add template DNAs (or PCR-grade water for no-template controls; *see* **Note 11**). Always keep the tubes at 4 °C (on ice or in a cold block) prior to thermal cycling, in order to minimize unspecific amplification that may arise from residual *Taq* polymerase activity at low temperatures.

7. Close the strip tubes with their lids. Make sure that a tight seal is formed to prevent evaporation of the reaction mixes during thermal cycling.

8. Transfer the tubes into the thermal cycler and initiate the cycling protocol. Note the positions in which the tubes are placed within the thermal block (*see* **Note 14**).

9. At the end of the cycling protocol, remove the tubes from the thermal cycler. Store the products at 4 °C until further analysis or other downstream applications such as DNA sequencing.

3.3 Agarose Gel Electrophoresis

1. On an electronic scale, weigh out the required amount of agarose gel powder (*see* **Note 15**).

2. Add the agarose powder into an adequate volume of 1× TAE buffer in a glass bottle and swirl briefly to mix.

3. Heat the mixture in a microwave oven until it begins to boil. The bottle should only be loosely closed with a lid to avoid a dangerous buildup of pressure inside it.

4. Remove the bottle from the microwave oven and gently swirl the partially molten agarose mixture. Wear a pair of heatproof gloves and safety glasses while doing this.

5. Replace the bottle in the microwave and repeat **steps 3** and **4** twice or until the agarose powder completely dissolves in the buffer. At this point the cloudy mixture should turn into a clear solution.

6. Allow the agarose solution to stand on the bench or in a water bath until it has cooled to ~50 °C. Alternatively, hold the bottle under cold running tap water and gently swirl the molten agarose solution to make sure that it is thoroughly cooled. Do not over-cool as the agarose solution will begin to solidify (*see* **Note 16**).

7. Add an adequate amount of nucleic acid stain (*see* **Note 17**), swirl to mix, and set aside the stained agarose solution to cool further.

8. Place the gel tray, in its casting position, into the buffer chamber. Pour the warm molten agarose solution onto the gel tray and insert a well comb. Do not pour the agarose solution while it is still hot as this may cause the gel tray to warp.

9. Once the gel solidifies, remove the comb and add the TAE buffer until it covers the entire gel to a depth of about 1 mm. Make sure that all the wells are properly formed.

10. Prior to loading, mix the PCR products with an adequate amount of DNA loading dye on a piece of Parafilm® (Bemis Co., Inc., Oshkosh, WI, USA; *see* **Note 18**). This can be done by pipetting the mixture up and down. Do not move the buffer chamber after loading, as this may cause cross-contamination between samples.

11. Set the electrophoresis power supply to provide 5–8 V/cm of the distance between the anode and the cathode (typically ~80 V for a gel of 8 cm length). Electrophorese the PCR products for ~45 min or until the dye front has visibly traversed half the length of the gel. A DNA size marker should be run alongside the PCR products.

4   Notes

1. Commercially synthesized primers are typically lyophilized for ease of shipment. On receipt, the tubes should be briefly centrifuged prior to resuspension of the DNA pellet, as dried pellets may become dislodged during transport. Storage at −20 °C should extend the shelf-life of the primer stocks.

2. Use of diluted working solutions is advisable, in order to minimize the number of freeze-thaw cycles and to preserve the primary stock of primers in the event of contamination. Primer solutions are stable for up to six months when stored at 4 °C [12]. For a detailed guide on primer resuspension and storage, please *see* ref. 12.

3. Commercial preparations that have worked well in our laboratory are KAPA LongRange HotStart DNA Polymerase (Kapa Biosystems, Inc., Wilmington, MA, USA), PrimeSTAR® GXL DNA Polymerase (Takara Bio, Inc., Otsu, Shiga, Japan), and AtMax Taq DNA Polymerase (Vivantis Technologies Sdn. Bhd., Subang Jaya, Selangor, Malaysia). The recommended cycling protocol and the amplification limit vary between these enzyme systems; however, they were noted to be equally effective when trialled for *CYP2D6* sequence analysis. Other polymerase formulations are likely to be suitable, but it is wise to test more than one for specific applications.

4. When not regularly used these reagents should be stored at −20 °C; otherwise they can be kept at 4 °C.

5. Input DNA should be checked for purity and integrity prior to PCR. An OD_{260}/OD_{280} (optical density) ratio of ~1.8 indicates pure DNA [13]. DNA should be of high molecular weight [5], preferably exceeding 20–40 kb. DNA templates of lower molecular weight may result in substantially decreased amplification efficiency. For routine use in PCR, a diluted aliquot of a DNA sample should be taken and kept at 4 °C; the remainder of the sample should be stored at −20°C [14].

6. Betaine and DMSO may sometimes be required for the amplification of GC-rich regions [15]. For instance, we routinely add betaine, at a final of concentration of 1 M, into the long PCR assays employed for *CYP2D6* sequence analysis [16].

7. The buffer should be made up in a fume cupboard, as glacial acetic acid is corrosive and has a strong, unpleasant odor. In addition to such general laboratory precautions, double gloving may be necessary for handling the chemical.

8. Melting temperatures are predicted by the nearest-neighbor formula with salt correction, assuming the concentrations of monovalent cations, divalent cations, oligonucleotides, and dNTPs to be 50 mM, 1.5 mM, 0.5 μM, and 0.6 mM, respectively [17]. For further details about how these concentrations are derived, please see *Advanced Parameters* in Primer-BLAST.

9. A PCR protocol for the amplification of a 6.6-kb fragment from the *CYP2D6* gene is shown here. DNA samples are initially denatured at 94 °C for 3 min and then subjected to thermal cycling as follows: 35 cycles of 94 °C for 25 s, 68 °C for 10 s, and 68 °C for 7 min, and a final elongation step of 72 °C for 7 min. For the amplification of very long products exceeding

15 kb in size, a two-phase auto-extension protocol may be adopted. The first phase comprises ten cycles of usual steps of heating and cooling. In the second phase, the extension time is successively prolonged by 20 s per cycle, for a total of 25 cycles. The additional extension time ensures that the synthesis of PCR products progresses to completion in each cycle. Addition of DNA ligase may further augment the efficacy of the PCR protocol [18], though we have not found such benefit to be consistently reproducible.

10. Long-range PCRs may be sensitive to the characteristics of a thermal cycler. This could be due to two reasons: (a) the well temperatures are not consistent across different Peltier heating blocks or different machines and (b) the thermal ramp rates vary between thermal cyclers.

11. Calculate the volumes of all reagents required for $n + 1$ samples to account for manipulative loss. Always include a no-template control for each primer pair. A typical reaction may comprise 1.5–2 mM of magnesium chloride, 0.2–0.5 µM of forward or reverse primer, 0.1–0.2 mM of each nucleotide, and 1.25–2.5 U of *Taq* polymerase; though the composition may vary with the enzyme system used and the target region. The enzymes should be added last into the reaction mixture, especially when this already contains a DNA template, to minimize unspecific amplification at low temperatures. For a 10-µL reaction, 30–50 ng of DNA is usually sufficient for ample amplification. Adding too much DNA may result in PCR inhibition (Fig. 3).

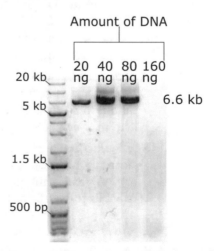

Fig. 3 The effect of different amounts of input DNA on the *CYP2D6* long-range PCR, performed in 10-µL aliquots, was ascertained. It is obvious that 4 ng/µL of DNA resulted in greater amplification than 2 ng/µL. Increasing the concentration to 8 ng/µL did not enhance the yield further; doubling it to 16 ng/µL was clearly detrimental. The samples used in these experiments were human genomic DNAs

12. The reaction volume may be adjusted depending on the downstream applications. For instance, if the long amplicons serve as sequencing templates in subsequent procedures, a large reaction volume, ranging from 25 to 50 µL, may be required.

13. Colorless PCR strip tubes are preferred. We have found some colored PCR tubes to have an inhibitory effect on long-range PCRs, causing amplification failure [19]. When there is a sudden decline in PCR performance, it may be worthwhile to check the PCR tubes. If these are identified as the source of PCR inhibition, the following strategies may be effective in remedying the problem: (a) use a different batch of the same tubes; (b) try PCR tubes from a different supplier; or (c) where colored PCR tubes are probably the culprit, switch to their colorless counterparts.

14. The effect of heating may not be entirely consistent across the assemblage of wells in a thermal cycler. Such positional effect may sometimes affect the performance of long-range PCRs and can be minimized by regular calibration of the thermal cycler.

15. Resolution of fragments larger than 1 kb typically requires the use of gels with concentrations ranging from 0.7 to 1%. For a 1% gel measuring 7 × 8 cm (width × length), add 0.5 g of agarose powder in 50 mL of TAE buffer.

16. Where an agarose gel has not homogenously formed, the samples loaded within different wells may not migrate in a synchronous manner.

17. For a 10,000× DNA stain concentrate, add 5 µL of the stain in 50 mL of TAE buffer. If ethidium bromide is used, strict disposal procedures should be followed; this is not necessary for other safer alternatives such as SYBR® Safe (Invitrogen™, Life Technologies, Carlsbad, CA, USA).

18. Usually 10–20% of PCR products should be sufficient for visualization, but more could be loaded if the products have no other use after electrophoresis. For instance, 2–4 µL of PCR products may be loaded if the starting reaction volume is 20 µL.

References

1. Saiki RK, Gelfand DH, Stoffel S, Scharf SJ, Higuchi R, Horn GT et al (1988) Primer-directed enzymatic amplification of DNA with a thermostable DNA polymerase. Science 239(4839):487–491

2. Ou CY, Kwok S, Mitchell SW, Mack DH, Sninsky JJ, Krebs JW et al (1988) DNA amplification for direct detection of HIV-1 in DNA of peripheral blood mononuclear cells. Science 239(4837):295–297

3. Barnes WM (1994) PCR amplification of up to 35-kb DNA with high fidelity and high yield from lambda bacteriophage templates. Proc Natl Acad Sci U S A 91(6):2216–2220

4. Cheng S, Fockler C, Barnes WM, Higuchi R (1994) Effective amplification of long targets from cloned inserts and human genomic DNA. Proc Natl Acad Sci U S A 91(12):5695–5699

5. Cheng S, Chen Y, Monforte JA, Higuchi R, Higuchi R, Van Houten B (1995) Template integrity is essential for PCR amplification of 20- to 30-kb sequences from genomic DNA. PCR Methods Appl 4(5):294–298

6. Fukami T, Nakajima M, Sakai H, McLeod HL, Yokoi T (2006) *CYP2A7* polymorphic alleles confound the genotyping of *CYP2A6*4A* allele. Pharmacogenomics J 6(6):401–412

7. National Center for Biotechnology Information (2013) The NCBI handbook [Internet], 2nd edn. National Center for Biotechnology Information, Bethesda, MD. http://www.ncbi.nlm.nih.gov/books/NBK143764/

8. Untergasser A, Cutcutache I, Koressaar T, Ye J, Faircloth BC, Remm M et al (2012) Primer3 – new capabilities and interfaces. Nucleic Acids Res 40(15):e115

9. Koressaar T, Remm M (2007) Enhancements and modifications of primer design program Primer3. Bioinformatics 23(10):1289–1291

10. Cone RW, Fairfax MR (1993) Protocol for ultraviolet irradiation of surfaces to reduce PCR contamination. PCR Methods Appl 3(3):S15–S17

11. Fox JC, Ait-Khaled M, Webster A, Emery VC (1991) Eliminating PCR contamination: is UV irradiation the answer? J Virol Method 33(3):375–382

12. Devor EJ, Behlke MA (2004) Oligonucleotide yield, resuspension, and storage. Integrated DNA technologies. https://www.idtdna.com/pages/decoded/decoded-articles/core-concepts/decoded/2012/10/08/dna-oligonucleotide-resuspension-and-storage. Accessed 15 Jun 2016.

13. Barbas CF, Burton DR, Scott JK, Silverman GJ (2007) Quantitation of DNA and RNA. CSH Protoc. doi:10.1101/pdb.ip47

14. Röder B, Frühwirth K, Vogl C, Wagner M, Rossmanith P (2010) Impact of long-term storage on stability of standard DNA for nucleic acid-based methods. J Clin Microbiol 48(11):4260–4262

15. Frackman S, Kobs G, Simpson D, Storts D (1998) Betaine and DMSO: enhancing agents for PCR. Promega Notes 65(27-29):27–29

16. Wright GE, Niehaus DJ, Drögemöller BI, Koen L, Gaedigk A, Warnich L (2010) Elucidation of *CYP2D6* genetic diversity in a unique African population: implications for the future application of pharmacogenetics in the Xhosa population. Ann Hum Genet 74(4):340–350

17. SantaLucia J Jr (1998) A unified view of polymer, dumbbell, and oligonucleotide DNA nearest-neighbor thermodynamics. Proc Natl Acad Sci U S A 95(4):1460–1465

18. Ignatov KB, Kramarov VM (2009) DNA ligases from thermophilic bacteria enhance PCR amplification of long DNA sequences. Biochemistry (Mosc) 74(5):557–561

19. Chua EW, Miller AL, Kennedy MA (2015) Choice of PCR microtube can impact on the success of long-range PCRs. Anal Biochem 477:115–117

Chapter 4

Strategies to Improve Efficiency and Specificity of Degenerate Primers in PCR

Maria Jorge Campos and Alberto Quesada

Abstract

PCR with degenerate primers can be used to identify the coding sequence of an unknown protein or to detect a genetic variant within a gene family. These primers, which are complex mixtures of slightly different oligonucleotide sequences, can be optimized to increase the efficiency and/or specificity of PCR in the amplification of a sequence of interest by the introduction of mismatches with the target sequence and balancing their position toward the primers 5′- or 3′-ends. In this work, we explain in detail examples of rational design of primers in two different applications, including the use of specific determinants at the 3′-end, to: (1) improve PCR efficiency with coding sequences for members of a protein family by fully degeneration at a core box of conserved genetic information, with the reduction of degeneration at the 5′-end, and (2) optimize specificity of allelic discrimination of closely related orthologous by 5′-end degenerate primers.

Key words PCR, Degenerate primers, 5′-End, 3′-End, Specificity determinants, PCR efficiency, MAMA-DEG PCR

1 Introduction

PCR or polymerase chain reaction is a molecular biology technique first developed by Kary B. Mullis that allows the amplification of a segment of DNA in theory from as little as a single DNA molecule [1]. To carry out a PCR reaction and produce double-stranded DNA molecules, the four building blocks of the DNA (the deoxynucleotides or dNTPs) must be present together with oligonucleotides (also called mers or primers), a DNA polymerase, and a DNA template chain. The use of thermo-resistant enzymes enable the production of large amounts of polymers of DNA, flanked by primer sequences, after repetition of cycles of high and low temperature that denature the double-stranded DNA chain, and allow the annealing of the primers to the template DNA and the DNA

Lucília Domingues (ed.), *PCR: Methods and Protocols*, Methods in Molecular Biology, vol. 1620,
DOI 10.1007/978-1-4939-7060-5_4, © Springer Science+Business Media LLC 2017

extension [2]. Primers must be complementary to the upper and lower chains and can be easily designed when the target DNA sequence is well known. However, sometimes the DNA target sequence is unknown, or presents variability. Usually, when a DNA sequence is unknown for one organism, its protein sequence might be deduced from orthologous that belong to the same protein family, since the conservation of structure–function relationships relies on amino acid sequence homology [3]. In these cases, it is possible to infer the DNA target sequence through the reverse translation of the amino acids sequence to the coding DNA, although the redundancy of the genetic code imposes a degree of uncertainty on the DNA sequence. This can be overcome by producing sequences of primers, called degenerate, made of a collection of sequences instead of one single sequence as in specific primers [4]. The International Union of Pure and Applied Chemistry (IUPAC) have established a nomenclature for incompletely specified bases in nucleic acid sequences [5] (Table 1). Since only a few sequences within a degenerate oligonucleotide might be functional for every particular DNA template, strategies to reduce its complexity are strongly recommended in order to set an optimal PCR efficiency. Different roles for the 5′- and 3′-ends of primer molecules can be predicted. Thus, whereas the 5′-end contributes marginally, match of the 3′-end is critical for PCR efficiency during DNA synthesis, since it is where the 3′-OH substrate needed for DNA polymerase activity is located in a double helix template [6]. It has been shown that one single mismatch in the last three nucleotides counting from the 3′-end is acceptable, but a second impaired position drops efficiency bellow the detectable level on agarose gels electrophoresis [7].

2   Materials

2.1 Sequence Manipulation and Primer Design

1. Text processor, e.g., Microsoft Word, OpenOffice Text Document or Pages for Mac.

2. Multiple alignment programs, e.g., http://www.ebi.ac.uk/Tools/msa/ or http://embnet.vital-it.ch/software/ClustalW.html.

3. Primer design algorithms, e.g., http://www.ncbi.nlm.nih.gov/tools/primer-blast/, http://simgene.com/Primer3 or https://tools.thermofisher.com/content.cfm?pageid=9716&icid=fr-oligo-6?CID=fl-oligoperfect or commercial informatics program as Oligo Primer Analysis Software (Molecular Biology Insight).

Table 1

Genetic code and compressed notation code according to IUPAC [5]

Amino acid		Codons	Compressed notation codon	Nucleotide	Base
A	Alanine	GCT, GCC, GCA, GCG	GCN	A	Adenine
R	Arginine	CGT, CGC, CGA, CGG, AGA, AGG	CGN, MGR	C	Cytosine
N	Asparagine	AAT, AAC	AAY	G	Guanine
D	Aspartic acid	GAT, GAC	GAY	T	Thymine
C	Cysteine	TGT, TGC	TGY	R	A or G
Q	Glutamine	CAA, CAG	CAR	Y	C or T
E	Glutamic acid	GAA, GAG	GAR	S	G or C
G	Glycine	GGT, GGC, GGA, GGG	GGN	W	A or T
H	Histidine	CAT, CAC	CAY	K	G or T
I	Isoleucine	ATT, ATC, ATA	ATH	M	A or C
L	Leucine	TTA, TTG, CTT, CTC, CTA, CTG	YTR, CTN	B	C or G or T
K	Lysine	AAA, AAG	AAR	D	A or G or T
M	Methionine	ATG		H	A or C or T
F	Phenylalanine	TTT, TTC	TTY	V	A or C or G
P	Proline	CCT, CCC, CCA, CCG	CCN	N	Any base
S	Serine	TCT, TCC, TCA, TCG, AGT, AGC	TCN, AGY		
T	Threonine	ACT, ACC, ACA, ACG	ACN		
W	Tryptophan	TGG			
Y	Tyrosine	TAT, TAC	TAY		
V	Valine	GTT, GTC, GTA, GTG	GTN		
Start		ATG			
Stop		TAA, TGA, TAG	TAR, TRA		

2.2 PCR, Gel Electrophoresis and Image Acquisition

1. 200 µl and 1.5 ml DNase-free plastic microtubes tubes (*see* **Note 1**).

2. Termocycler.

3. Gel electrophoresis system.

4. UV transiluminator or gel acquisition system.

2.3 Reagents

1. DreamTaq green DNA polymerase (Thermo Fisher) and buffer (*see* **Note 2**).

2. dNPTs (*see* **Note 3**).

3. Degenerate primers.

4. Nuclease-free water or autoclaved to sterility ultrapure water obtained by a water purification system.

5. DNA stains, e.g., GelRed (Biotium).

6. Gel-electophoresis agarose.

7. DNA molecular weight marker, e.g., GeneRuler 1 kb Plus DNA ladder (Thermo Scientific).

8. Gel electrophoresis Buffer TAE (*see* **Note 4**).

3 Methods

3.1 Guidelines for Primer Design from Protein Family Alignment: Reduction of 5′-End Complexity

Fully conserved motifs of 4–6 consecutive amino acids, the core box, are required to design the oligonucleotide 3′-end with, at least, 11 nucleotide positions. This is considering that the first position, in the reverse primer, or the last position, in the forward primer, leading to ambiguity is omitted from the oligonucleotide sequence to start reducing its complexity. In general terms, a good principle for designing highly efficient degenerate primers would be to establish a limit of degeneracy up to 64 (the number of different sequences within the mixture primer), meaning that the core box should not contain at all amino acid residues encoded by six codons (L, R, or S) and very few encoded by 3 (I) or four codons (A, G, P, V, or T). In contrast, amino acids encoded by two (C, D, E, F, H, K, N, T, or Y) or only one codon (M or W) are strongly recommended (Table 1). The remaining sequence of primers, typically 7–12 nucleotides toward the 5′-end, are less critical for DNA synthesis and, thus, is the region where efforts to reduce degeneration should be focused. An example of core box selection for primer design is shown in Fig. 1, where relationships between the two class A β-lactamases found in *Bacteroides* and related microorganisms, encoded by *cepA* and *cfxA* genes, are shown. Two primers were designed for *cepA* selective PCR from core boxes containing 3′-end specificity determinants (Table 2), allowing genotyping of β-lactam resistant strains of gram-negative anaerobes isolated from human and animals [8, 9].

The strategies to reduce primer complexity at the 5′-end include consideration of the codon usage bias, which is particular for every organism [10]. So when the core box is extended toward the 5′-end of the oligonucleotide, the information contained in the protein sequence might be reverse translated by using the genetic code and the particular codon usage bias of the target

```
                                    CepA1                    CepA2
Bfra    110...PQGGIEMSIADLLKYTLQQSDNNACDI......92......VTMGHKTGTGDRNAKG...55
Bun1    109...PDQDFTITLRELMQYSISQSDNNACDI......92......TVVGHKTGSSDRNADG...53
Bthe    105...PQGGFNIDIADLLNYTLQQSDNNACDI......92......VTIGHKTGTGDRNAKG...53
Bun2    127...SGPVISLTVRDLLRYTLTQSDNNASNL......94......VVIAHKTGSGYVNENG...57
Pden    127...SGPVISLTVRDLLRYTLTQSDNNASNL......94......VVIAHKTGSGDVNENG...57
Coch    127...SGPVISLTVRDLLRYTLTQSDNNASNL......94......VVIAHKTGSGDVNENG...57
             .      : :  :  :*:.*:: ******.::      ...:.****:.  *  .*
```

Fig. 1 Protein alignment and core box selection for primer design. Sequences shown are: Bfra, AAA22905.1; Bun1, AAA66962; Bthe, NP_813418.1; Bun2, AAB17891; Pden, AAM48119.1; Coch, AAL79549.2. Multiple alignement was performed by Clustal Omega. *Bold characters* represent strongly conserved positions. *Grey boxes* indicate identical residues. *Black boxes* represent CepA characteristic residues used to design primers with 3′-end specific determinants for discriminatory PCR. The expected size of *cepA* fragments amplified by PCR with cepA1/A2 primers is 329 bp [8]

Table 2

Primer design for detection of *cepA* orthologous from *Bacteroides* and related anaerobic gram-negative bacteria

Forward primer							
Amino acids	S	D	N	N	A	C	D
Coding sequence	TCN AGC T	GAC T	AAC T	AAC T	GCN	TGC T	GAC T
Abbreviated	WSN	GAY	AAY	AAY	GCN	TGY	GAY
	↓						↓
CepA1 (×64)	AGC	GAY	AAY	AAY	GCN	TGY	GA
Reverse primer							
Amino acids	G	H	K	T	G	T/S	G/S
Coding sequence	GGN	CAC T	AAA G	ACN	GGN	ACN TCN AGC T	GGN TCN AGC T
Abbreviated	GGN	CAY	AAR	ACN	GGN	WSN	BSN
	←						→
Complementary	NSH	NSW	NCC	NGT	YTT	RTG	NCC
CepA2 (×64)	GA	AGA	TCC	NGT	YTT	RTG	NCC

organism, reducing to a minimum the degeneration of its DNA sequence.

Another possible approach would be to combine the reverse translation of the core box and the multiple alignment of DNA sequences encoding homologue sequences. The 5'-end sequence of primers could be assumed directly from the alignment, considering the organism codon usage, and further, the particular evolution of DNA sequences within the gene family. However, when the genetic variability extend toward the 5'-end of oligonucleotides and there is no rational way to reduce primer degeneracy to a value lower than 64, or in extreme circumstances to 128 (very few published reports have signaled success by using primer degeneracy higher than 128), we recommend the design of different primers sequence variants that might be used separately in different PCR. This strategy should result better than using inosine phosphate as a degenerate base, since although this permissive nucleotide can base-pair with any nucleotide without increasing sequence complexity [11], in our experience its presence confers low reliability to PCR primers. The core boxes selected for oligonucleotide design shown in Fig. 1 constitute an example of managing the 5'-end of degenerate oligonucleotides to reduce their complexity. Whereas reverse primer lacks strongly specific 3'-end (only one position is fully variable among homologous), a fully specific PCR was expected to be determined by the forward primer (Fig. 1). The fully degenerated 3'-ends deduced from both core boxes present the complexity threshold for efficient PCR (64) from 6 amino acid residues for the forward primer (17 nucleotides) and from 5 residues for the reverse primer (15 nucleotides), which are nearly 3/4 of a typical 21-bases oligonucleotides. Since specificity and efficiency requirements were fulfilled, the remaining sequences toward the 5'-ends were taken randomly from a DNA sequence alignment (not shown) to balance the GC content near 50%, which is also recommended.

3.2 Guidelines for Primer Design from DNA Alignment: Reduction of Specificity by Degeneration of 5'-End

In primer design for allelic discrimination, the number of mismatches at the last positions from the 3'-end of the primer is crucial for annealing and specific elongation by the DNA polymerase, which supports the mismatch amplification mutation assay or MAMA-PCR [7]. The discrimination power of this technique is based in the complementarity of each particular allele of a polymorphic position to the 3'-end of their corresponding primers, whose annealing is weakened by an additional mismatch with their target sequences, in a way that allows amplification with one but not two mismatches. The technique is so sensitive that additional mispairing of primers produced by genetic variability of sequences would give rise to false negatives. To solve that, in a particular case that requires the use of DNA from closely related species as target

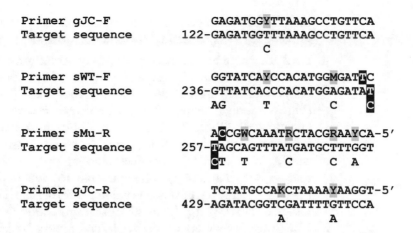

Fig. 2 *gyrA* sequence variability in human thermophilic *Campylobacter*: primer design for discrimination the C-257-T polymorphism. Primers are shown above their target sequence, the *gyrA* gene from *C. jejuni*. Residues in *grey boxes* represent primers degenerate positions (IUPAC code), according to the polymorphism of *gyrA* that exists among described sequences from *C. jejuni* and *C. coli* that is shown below the target sequence. Residues in *black boxes* are critical for MAMA-PCR. In the target sequence they represent the allele-specific polymorphism whereas in primer sequence indicate the mismatch that weakens the annealing and enable discriminatory PCR. The expected product sizes, for multiple PCR with the four primers, are 157 plus 329 bp, for T-257 *gyrA* linked to quinolone resistant isolates, or 215 plus 329 bp, for C-257 *gyrA* of quinolone sensitive *Campylobacter* [12]

sequences, we designed degenerate primers with allele-specific determinant at their 3′-ends that allow discrimination of the C-257-T polymorphism of *gyrA* sequence, responsible for quinolone resistance of *Campylobacter* isolates [12]. This primer design, shown in Fig. 2, was performed by including degeneration toward the primer 5′-ends, increasing the target recognition to both major species of thermophilic *Campylobacter* isolated from humans, *C. jejuni* and *C. coli*, without compromising the allele specific amplification of *gyrA* for both species.

3.3 Protocol for Protein or DNA Sequence Analysis

1. Gather all the DNA or protein sequences of interest from online databases as the National Center for Biotechnology Information (NCBI, http://www.ncbi.nlm.nih.gov/nuccore for nucleotides or http://www.ncbi.nlm.nih.gov/protein/ for proteins) in FASTA format (*see* **Note 5**).

2. Paste the sequences in a text processor document.

3. Copy all the sequences and use the algorithm Clustal on online platforms to align the multiple DNA or protein sequences. Clustal can be found in the EBI multi Sequence Alignment (http://www.ebi.ac.uk/Tools/msa/), the Swiss Institute of Bioinformatics (www.ch.embnet.org/software/ClustalW.html),

the Kyoto University Bioinformatics Center (http://www.genome.jp/tools/clustalw/) or other platforms.

4. Look for a core box of successive and conserved amino acid residues in the protein alignment (Fig. 1) or a 18 to 24 conserved DNA sequence around the target polymorphism (*see* **Note 6**) (Fig. 2). Protein sequences must be reverse translated to their genetic code (Table 2). This can be done on platforms as the Sequence Manipulation Suit (http://www.bioinformatics.org/sms2/) or BioPHP (http://www.biophp.org/minitools/protein_to_dna/demo.php). It is possible to obtain the codon usage of several organisms on online platforms like The Codon Usage Database (http://www.kazusa.or.jp/codon/) or EMBL-EBI (http://www.ebi.ac.uk/Tools/st/emboss_backtranseq/).

5. Consider a conserved region that has at the penultimate or ultimate position of the 3′ end of the possible primer a conserved position that can be used to distinguish a desirable trait and design the primer considering degenerate positions between target sequences, preferably at the primer 5′-ends (Table 1). An extra and obligated mismatch at the 3′-end should be included to weaken primer annealing and increase specificity toward allele-specific (Fig. 2).

6. Primer design quality can be assessed by checking the formation of hairpins and primer dimerization.

7. Once primers are designed they can be ordered using an online service from an oligonucleotide synthesis company.

3.4 PCR Reaction

A successful PCR reaction relies on a number of factors, some of which related to the quality of the primer design. Primer length should be between 16 and 28 nucleotides, which depending on the GC content should produce primers' melting temperatures (Tm) between 50 and 62 °C. It is not judicious to have a Tm difference between the two primers greater than 5 °C [13], and the annealing temperature can be empirically determined as being 5 °C lower than the primer with the lowest Tm. Since for degenerate primers the working sequence(s) of primers is(are) unknown, a good approach is to set annealing temperature between 50 and 55 °C, although specific primers corresponding to one known homologue could be used to estimate more accurately the PCR conditions.

Other important aspect in the success of the PCR reaction is the quality of the genomic DNA that includes the target sequence. Due to contamination problems, isolated areas in the laboratory should be dedicated to the DNA extraction process and PCR examination area, where gels are run, whilst PCR master mixes and all PCR reagents should be handle in a laminar flow cabinet.

1. Bacterial DNA suitable for PCR can be obtained by the boiling method (*see* **Note 7**) or by using any commercial DNA extraction kit.

2. Prepare a PCR master mix. For a PCR reaction with a final volume of 20 μl mix in a 1.5 ml microtube the following (*see* **Note 8**):

(a) 1–1000 ng of genomic DNA or 1–1000 pg of plasmid DNA (usually 1–2 μl of DNA),

(b) 2 μl of *Taq* polymerase buffer (usually comes 10x concentrated and is used 1x concentrated),

(c) 0.4 μl of dNTPs solution with the concentration of 10 μM,

(d) 1 μl of forward primer solution with the concentration of 10 mM,

(e) 1 μl of reverse primer solution with the concentration of 10 mM,

(f) 0.2 μl *Taq* polymerase at the concentration of 5 U/μl,

(g) Water up to the volume of 20 μl.

3. Mix gently.

4. Add to each 200 μl PCR microtubes the sample DNA plus the amount of master mix to obtain a final volume of 20 μl.

3.5 Thermocycling

Incubate the 200 μl PCR microtubes containing the DNA and the master mix in a thermocycler with the following program: initial denaturation 95 °C for 2 min, 35 cycles of denaturation at 95 °C for 30 s, primer annealing between 50 °C and 55 °C for 1 min, extension at 72 °C for 1 min, a final extension at 72 °C for 10 min. The PCR can be maintained at room temperature indefinitely.

3.6 Gel Analysis

1. Prepare a 0.75 to 1.0% agarose gel (*see* **Note 9**).

2. Load 10 μl of each PCR product in each gel well.

3. Connect the power supply to 80 V and wait approximately 1 h or until the fastest moving band in the PCR buffer is close to the gel hedge.

4. Observe the gel in the UV transiluminator or gel acquisition system and look for the presence of the expected size band.

4 Notes

1. If tubes are not DNase free, sterilization at 121 °C for 15 min is suitable for DNase elimination.

2. DreamTaq green buffer allows direct loading of PCR products in gel wells since it has a density reagent and two tracking dyes.

3. dNTPs are available as mixtures of all the dNTPs or as single solutions of dATP, dCTP, dGTP, and dTTP.

4. TAE buffer is commercialized concentrated 50× or can be prepared with the following composition: 40 mM Tris (pH 7.6), 20 mM acetic acid, and 1 mM EDTA. Should be used 1× concentrated for gel preparation and in electrophoresis systems.

5. Nucleotide or protein sequences can be obtained in the FAST format. This format is widely used in bioinformatics with each dNTP being represented with the letters A, T, C and G and each amino acid represented by the IUPC single letter code [5]. Each FASTA file begins with the symbol ">" followed by a description line which can be used as an identifier of the sequence. The NCBI FASTA format recommends lines of text to be shorter than 80 characters in length (http://blast.ncbi. nlm.nih.gov/blastcgihelp.shtml).

6. At the bottom of a DNA sequence alignment the symbol "*" indicates a perfect alignment and a lack of symbol represents a nonconserved sequence. In protein sequence alignments two other symbols are presented. ":" Indicates a position with conserved amino acids with strong similarity, whereas "." indicates a position with conserved amino acids with weak similarity.

7. To use the boiling method, bacterial cells should be grown on solid media. One isolated colony is suspended in 200 µl of sterile ultrapure water or sterile TE buffer in a microtube. The tube is heated in a thermo block at 100 °C for 10 min. After that the tube is centrifuged at $10,000 \times g$ for 5 min. The supernatant can be directly used in PCR reactions and can be stored at −20 °C for further use.

8. Multiple the amounts of each reagent by the number of samples plus two. This will guarantee enough reaction volume, since handling loss may occur. Make sure to include one negative control (water instead of DNA) and one positive control (a sample known to be positive) in your samples. All reagents and tubes should be kept on ice until placed in the thermocycler to guarantee the polymerase activity.

9. Big amplicons are better resolved in lower concentration of agarose gels (0.7–1%), while small amplicons are better resolved in higher concentrations of agarose gels (1–1.5%).

Acknowledgments

This study had the support of Fundação para a Ciência e Tecnologia (FCT), through the strategic project UID/MAR/04292/2013 granted to MARE, the Department of Economy and Infrastructure of the Regional Government of Extremadura (Group CTS001) and the University of Extremadura (Group MIVET).

References

1. Saiki RK, Scharf S, Faloona F, Mullis KB, Horn GT, Erlich HA, Arnheim N (1985) Enzymatic amplification of beta-globin genomic sequences and restriction site analysis for diagnosis of sickle cell anemia. Science 4732:1350–1354

2. Saiki RK, Gelfand DH, Stoffel S, Scharf SJ, Higuchi R, Horn GT, Mullis KB, Erlich HA (1988) Primer-directed enzymatic amplification of DNA with a thermostable DNA polymerase. Science 4839:487–491

3. Chothia C, Lesk AM (1986) The relation between the divergence of sequence and structure in proteins. EMBO J 4:823–826

4. Iserte JA, Stephan BI, Goñi SE, Borio CS, Ghiringhelli PD, Lozano ME (2013) Family-specific degenerate primer design: a tool to design consensus degenerated oligonucleotides. Biotechnol Res Int 2013:383646

5. Nomenclature Committee of the International Union of Biochemistry (NC-IUB) (1985) Nomenclature for incompletely specified bases in nucleic acid sequences. Recommendations. Biochem J 229(2):281–286

6. Haras D, Amoros JP (1994) Polymerase chain reaction, cold probes and clinical diagnosis. Sante 4(1):43–52

7. Cha RS, Zarbl H, Keohavong P, Thilly WG (1992) Mismatch amplification mutation assay (MAMA): application to the c-H-ras gene. PCR Methods Appl 2(1):14–20

8. García N, Gutiérrez G, Lorenzo M, García JE, Píriz S, Quesada A (2008) Genetic determinants for cfxA expression in *Bacteroides* strains isolated from human infections. J Antimicrob Chemother 5:942–947

9. Lorenzo M, García N, Ayala JA, Vadillo S, Píriz S, Quesada A (2012) Antimicrobial resistance determinants among anaerobic bacteria isolated from footrot. Vet Microbiol 157(1–2):112–118

10. Ikemura T (1985) Codon usage and tRNA content in unicellular and multicellular organisms. Mol Biol Evol 1:13–34

11. Ben-Dov E, Shapiro OH, Siboni N, Kushmaro A (2006) Advantage of using inosine at the 3′ termini of 16S rRNA gene universal primers for the study of microbial diversity. Appl Environ Microbiol 11:6902–6906

12. Hormeño L, Palomo G, Ugarte-Ruiz M, Porrero MC, Borge C, Vadillo S, Píriz S, Domínguez L, Campos MJ, Quesada A (2016) Identification of the main quinolone resistance determinant in *Campylobacter jejuni* and *Campylobacter coli* by MAMA-DEG PCR. Diagn Microbiol Infect Dis 3:236–239

13. Chuang LY, Cheng YH, Yang CH (2013) Specific primer design for the polymerase chain reaction. Biotechnol Lett 10:1541–1549

Chapter 5

Inverse PCR for Point Mutation Introduction

Diogo Silva, Gustavo Santos, Mário Barroca, and Tony Collins

Abstract

Inverse PCR is a powerful tool for the rapid introduction of desired mutations at desired positions in a circular double-stranded DNA sequence. Here, custom-designed mutant primers oriented in the inverse direction are used to amplify the entire circular template with incorporation of the required mutation(s). By careful primer design it can be used to perform such diverse modifications as the introduction of point mutations and multiple mutations, the insertion of new sequences, and even sequence deletions. Three primer formats are commonly used; nonoverlapping, partially overlapping and fully overlapping primers, and here we describe the use of nonoverlapping primers for introduction of a point mutation. Use of such a primer setup in the PCR reaction, with one of the primers containing the desired mismatch mutation, results in the amplification of a linear, double-stranded, mutated product. Methylated template DNA is removed from the nonmethylated PCR product by *Dpn*I digestion and the PCR product is then phosphorylated by polynucleotide kinase treatment before being recircularized by ligation, and transformed to *E. coli*. This relatively simple site-directed mutagenesis procedure is of major importance in biology and biotechnology today where it is commonly employed for the study and engineering of DNA, RNA, and proteins.

Key words Site-directed mutagenesis, Inverse PCR, Nonoverlapping primers, Protein engineering

---

1 Introduction

Site-directed mutagenesis (SDM) is a powerful method for making targeted, predetermined changes in a DNA sequence. It is invaluable in molecular biology and protein engineering for investigating the role of specific nucleotides and amino acids and for engineering desired properties into protein, DNA, and RNA molecules [1–6]. The original, relatively inefficient, SDM methods based on primer extension with single stranded DNA templates [7–9], have evolved over the years, and have been supplanted by the plethora of versatile, highly efficient SDM methods available today. Indeed, currently, a large variety of specific, high-throughput, in vitro [10, 11] and in vivo [12, 13] techniques and manufactured kits with

---

All authors contributed equally to this work.

Lucília Domingues (ed.), *PCR: Methods and Protocols*, Methods in Molecular Biology, vol. 1620,
DOI 10.1007/978-1-4939-7060-5_5, © Springer Science+Business Media LLC 2017

efficiencies up to almost 100% for the site-specific mutation of almost any sequence are available.

The most commonly used in vitro SDM methods employ PCR and are based on either overlap extension PCR [11, 14] or inverse PCR (iPCR) [15, 16] as well as modifications and combinations of these. Overlap extension PCR is more appropriate for linear sequences and requires multiple rounds of PCR, whereas iPCR is designed for circular templates such as vector insert sequences and uses a simplified protocol necessitating only one PCR reaction. Inverse PCR was first reported in 1988 for the identification of flanking regions of a known DNA sequence [15, 16]. Its designation, inverse, comes from the fact that the primers are oriented in the reverse direction, facing "outwards," away from each other, in contrast to regular PCR where "in-facing" flanking primers are employed.

Nonoverlapping-, partially overlapping, or fully overlapping primers can be used for SDM by iPCR. Nonoverlapping, "back-to-back" primers (Fig. 1a) produce a linear mutated sequence which must then be recircularized before transformation to E. coli [17]. Partially overlapping primers (Fig. 1b) yield a product with short homologous ends which can be directly transformed for in vivo recombination in E. coli [18–20]. Completely overlapping, complementary, inverse primers (Fig. 1c) form part of the widely used QuikChange SDM kit (Strategene), but currently the exact mechanism of action of this is under discussion. It had previously been proposed to progress by linear amplification of template to give a circular product, and not by exponential amplification as for a true PCR chain-reaction [20]. However, recent evidence [21] suggests exponential amplification of a linear product with short homologous ends for recombination, as with partially overlapping primers. Use of partially or fully overlapping primers allows for a more simplified SDM procedure than with nonoverlapping primers, but

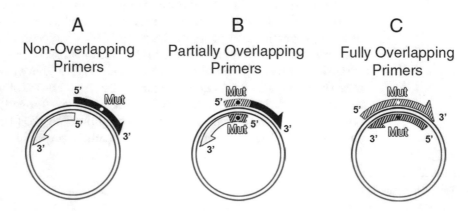

Fig. 1 Illustration of primer design formats for inverse PCR with nonoverlapping (**a**), partially overlapping (**b**) and fully overlapping (**c**) primers. The hatched sections show the overlapping regions of the primers

frequently necessitate longer more complex primers and are some-times characterized by poor or no amplification of PCR product, formation of primer dimers, and a reduced transformation efficiency [19–21].

In this chapter, we focuse on the iPCR method with nonoverlapping primers for introduction of a point mutation into a DNA sequence. This is composed of three principal steps (Fig. 2): (1) iPCR Mutant Amplification (including primer design, PCR and agarose gel confirmation), (2) Template Removal and Product Recircularization (including template digestion, mutant phosphorylation, and ligation), and (3) Transformation and Mutant Confirmation (transformation, plasmid construct isolation and sequencing). iPCR is relatively easy and rapid to employ and by simple modification of primer design, not only single base changes (point mutations), but also multiple base changes, deletions, and insertion can be carried out (see Fig. 2). Currently, a variety of optimized kits based on iPCR with nonoverlapping primers are available, e.g., the Phusion (Thermo Scientific), Q5 (New England Biolabs), and KOD-Plus (Toyobo) site-directed mutagenesis kits.

2 Materials

2.1 iPCR Mutant Amplification

1. PCR Thermal Cycler.

2. Thin-walled PCR tubes.

3. Circular, double-stranded, template DNA. Approximately 1 ng/µL stock, in autoclaved, ultrapure water is recommended, but may vary from 0.1 to 10 ng/µL depending on plasmid size, sequence, and quality (see Note 1).

4. PCR primers (see Note 2). For highest SDM efficiency, HPLC or PAGE purified primers are recommended. Resuspend lyophilized primers in 1.2 g/L (10 mM) Tris–HCl (tris(hydroxymethyl)aminomethane, adjust pH with HCl, pH 7) to a concentration of 100 µM and prepare 20 µM working stock solutions by dilution of aliquots in 1.2 g/L (10 mM) Tris–HCl, pH 7. All primer solutions should be stored at −20 °C and repeated freezing and thawing should be avoided.

5. High-fidelity DNA polymerase with proofreading activity (see Note 3), as supplied, e.g., Phusion High Fidelity DNA polymerase (2 U/µL). Store at −20 °C.

6. DNA Polymerase buffer (as supplied with the polymerase used), e.g., 5× concentrated Phusion HF buffer (see Note 4). Store at −20 °C.

7. Deoxyribonucleoside triphosphate (dNTP) mix, a stock solution of 10 mM is recommended. Store at −20 °C.

8. Autoclaved, ultrapure water (see Note 5).

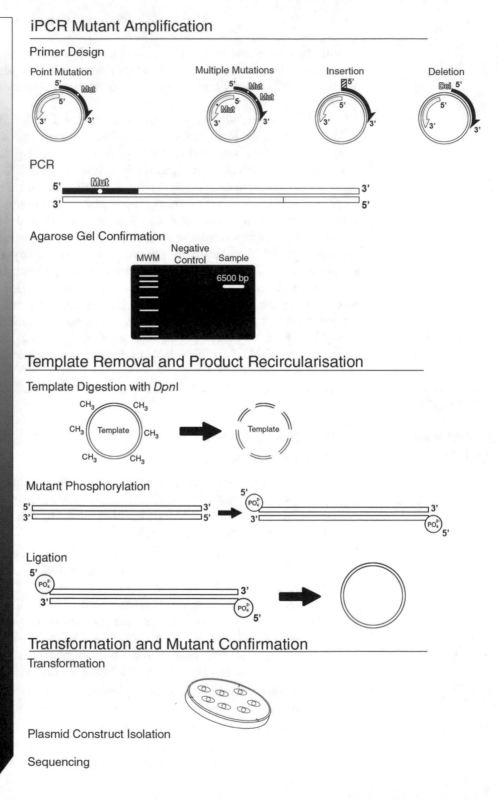

Fig. 2 Flowchart of the protocol for site-directed mutagenesis by inverse PCR with nonoverlapping primers. The primer design formats for introduction of a point mutation, multiple mutations, insertions, and deletions into a double-stranded circular DNA template are shown. Following primer design, the protocol employed for each type of mutation is identical. CH₃: methyl group of methylated DNA; PO_4^{2-}: phosphate group of 5′ phosphorylated DNA

2.2 Agarose Gel Electrophoresis

1. Agarose, molecular biology grade.

2. 50× (50 times concentrated) TAE solution (*see* **Note 6**): 242 g/L (2 M) Tris base, 60 g/L (1 M) glacial acetic acid, 14.6 g/L (50 mM) EDTA (Ethylenediaminetetraacetic acid). Store at room temperature. Dilute in deionized water to 1× (i.e., 50-fold dilution) prior to use.

3. 6× Loading buffer: 500 g/L glycerol, 58 g/L (0.2 M) EDTA, 0.5 g/L bromophenol blue, pH 8.3. Store at room temperature. Prepare 50 mL and store at room temperature for a maximum of 6 months (*see* **Note 7**).

4. Nucleic acid staining solution, e.g., Midori Green Advance (20,000×): 5 μL in 100 mL 1× TAE solution. While being significantly less mutagenic than the traditionally used ethidium bromide stain, appropriate care should be taken to avoid direct contact with midori green or similar nucleic acid stains. Store at −20 °C.

5. Molecular weight marker, as supplied. Store at −20 °C (*see* **Note 8**).

6. Gel casting trays and sample combs.

7. Electrophoresis chamber and power supply.

8. Transilluminator. Always use appropriate safety procedures and wear protective eyewear when using a transilluminator to prevent UV light damage.

2.3 Template Removal and Product Recircularization

1. Restriction enzyme *Dpn*I (as supplied, typically 10 U/μL). Store at −20 °C (*see* **Note 9**).

2. T4 DNA Ligase buffer (as supplied with T4 DNA Ligase, typically 10×, ensure this contains 5–10 mM ATP and 50–100 mM DTT) (*see* **Note 10**). Store at −20 °C.

3. T4 Polynucleotide Kinase (as supplied, typically 10 U/μL). Store at −20 °C (*see* **Note 11**).

4. T4 DNA Ligase (as supplied, typically 5 U/μL). Store at −20 °C (*see* **Note 12**).

5. Incubators at 37 and 25 °C (or room temperature).

2.4 Preparation of Chemically Competent E. coli XL1-Blue

1. *E. coli* XL1-Blue cells (*see* **Note 13**).

2. Transformation Buffer: 3 g/L (10 mM) PIPES, 1.7 g/L (15 mM) CaCl$_2$, 18.6 g/L (250 mM) KCl, 10.9 g/L (55 mM) MnCl$_2$·4H$_2$O. Mix all components except MnCl$_2$ and adjust pH to 6.7 with 112 g/L (2 M) KOH. Add MnCl$_2$ and mix, sterilize solution through a 0.4 μm membrane.

3. 100% dimethyl sulfoxide (DMSO).

4. SOB (Super Optimal Broth): 20 g/L bacto tryptone, 5 g/L yeast extract, 0.6 g/L (10 mM) NaCl, 0.75 g/L (2.5 mM) KCl, 0.95 g/L (10 mM) MgCl$_2$, 1.2 g/L (10 mM) MgSO$_4$.

2.5 Transformation

1. 200 μL aliquots of chemically competent *E. coli* XL1 Blue cells (*see* **Note 13**). Store at −70 °C.

2. pUC18 control plasmid (1 pg/μL). Store at −20 °C.

3. Luria Bertani Broth (LB): 10 g/L bacto tryptone, 5 g/L yeast extract, 10 g/L NaCl. Adjust pH to 7 with a 200 g/L (5M) NaOH solution. Autoclave to sterilize.

4. Ampicillin (*see* **Note 14**): 100 mg/mL stock solution in water. Filter-sterilize through a 0.4 μm membrane and store at 4 °C for no more than 1 month.

5. LB agar plates containing antibiotic (100 μg/mL ampicillin). Prepare LB as described above with addition of 18 g/L agar. Autoclave to sterilize, cool to 50–55 °C, add 1 mL/L of 100 mg/mL ampicillin stock, mix and aseptically pour to petri dishes.

2.6 Mutant Confirmation

1. LB (prepared as described above) + antibiotic (100 μg/mL ampicillin) (*see* **Note 14**).

2. 50 mL polypropylene falcon tubes.

3. Plasmid DNA purification kit, as supplied by manufacturer.

3 Methods

The protocol given is for the introduction of a single point mutation with Phusion high fidelity DNA polymerase in a 1007 bp sequence in the expression vector pET22b(+). The total construct size is 6500 bp and the selection marker is ampicillin resistance. Nevertheless, any templates up to ~10 kb in size and any rapid, high fidelity polymerase can be used with appropriate protocol modification according to manufacturer's recommendations (namely, modifications in the buffer and polymerase concentrations, the PCR cycle and/or antibiotic used) (*see* **Note 3**). PCR process optimization may be required in some cases.

3.1 iPCR Mutant Amplification

1. Primer Design. Inversed primers should anneal to opposite strands of the plasmid, be nonoverlapping and aligned back-to-back with apposing 5′ ends. Ideally, the targeted mismatch mutation should be located in the middle of the primer with 10–15 perfectly matched nucleotides on either side. Mutations can be incorporated closer to the 5′ end but at least ten complementary nucleotides are required at the 3′ end (*see* **Note 15**). Normal considerations for PCR primer design should be adhered to (*see* **Note 16**). Phosphorylation of primers is not required (*see* **Note 2**). For best results, at least HPLC grade purification of primers is required, for primers greater

than 40 nucleotides in length, PAGE purification is recommended (*see* **Note 2**).

2. PCR Reaction Setup (*see* **Note 17**). Add the following components in the order given to a thin walled PCR tube on ice: 13.4 µL autoclaved ultrapure water (for a total final reaction volume of 20 µL), 4 µL of 5× concentrated Phusion HF buffer (1× buffer) (*see* **Note 4**), 0.4 µL of 10 mM stock dNTP mix (200 µM of each dNTP), 0.5 µL of each 20 µM primer stock (0.5 µM of each primer) (*see* **Note 2**), 1 µL of 1 ng/µL plasmid template stock (1 ng) (*see* **Note 1**), and 0.2 µL of 2 U/µL Phusion DNA polymerase (0.4 U) (*see* **Note 3**). Gently mix, briefly centrifuge and immediately place in the thermal cycler. A negative control reaction with all components except the primers, which are substituted with an equal volume of water, should also be set up.

3. PCR Cycle (*see* **Note 18**): 1 cycle at 98 °C for 2 min, 25 cycles of denaturation, annealing and extension at, respectively, 98 °C for 20 s, the calculated primer annealing temperature for 20 s, and 72 °C for 2 min (~20 s/kb of template). A final extension is then carried out at 72 °C for 10 min before cooling to 4–10 °C. The same conditions are used for the sample and negative control.

4. Agarose Gel Confirmation. The results of the PCR are verified by visualizing 5 µL of the sample and negative controls on a 1% agarose gel using the following protocol. To 1 g of agarose add 1× TAE buffer to 100 mL and boil until the agarose is completely dissolved. When cooled to ~50–60 °C, pour into the casting tray, insert comb and leave until completely polymerized. Remove comb and place gel in the electrophoresis chamber, add 1× TAE buffer until gel is covered with solution. To 5 µL of sample and negative control, add 1 µL of 6× loading buffer, mix, and carefully pipette into agarose gel wells. Load molecular weight marker (*see* **Note 8**) into an adjacent well. Run the gel at 7–8 V/cm for 45–60 min, carefully remove and place in nucleic acid staining solution for 30 min before visualizing under a UV transilluminator. A strong band should be visible at 6500 bp for the sample and no band should be observed for the negative control (*see* **Note 19**).

3.2 Template Removal and Product Recircularization

1. Template Digestion. Add 1 µL of *Dpn*I directly to the PCR reaction (15 µL reaction volume remaining), mix by gently pipetting up and down, centrifuge briefly, and incubate for 30 min at 37 °C (*see* **Note 9**).

2. Phospho-ligation. To a new Eppendorf tube add 12 µL of autoclaved, ultrapure water, 2 µL of 10× T4 DNA ligase buffer

(*see* **Note 10**), 4 μL of *Dpn*I treated PCR product (*see* **Note 20**), 1 μL of PNK (10 U/μL) (*see* **Note 11**), and 1 μL of T4 DNA ligase (5 U/μL) (*see* **Note 12**). Mix by gently pipetting up and down, centrifuge briefly to spin down, and incubate for 90 min at 25 °C. Store on ice until transformation or store at −20 °C.

3.3 Transformation and Mutant Confirmation

1. Preparation of chemically competent *E. coli* XL1 Blue for transformation (*see* **Note 13**). Inoculate a single colony of *E. coli* XL1 Blue in 250 mL sterile SOB medium in a 2 L shake flask and incubate at 18 °C, 150 rpm until $OD_{600\ nm} = 0.6$. Place on ice for 10 min, centrifuge at 4 °C for 10 min at 2500 × g and decant the supernatant. Gently resuspend the pellet in 80 mL of ice-cold Transformation Buffer (care should be taken as cells are susceptible to mechanical disruption) and place on ice for 10 min. Centrifuge cells at 4 °C for 10 min at 2500 × g. Decant the supernatant and gently resuspend the pellet in 20 mL of ice-cold Transformation Buffer. Add DMSO to a final concentration of 7%, swirl gently and place on ice for 10 min. Dispense 200 μL aliquots in ice cold 1.5 mL Eppendorf tubes and immediately freeze in liquid nitrogen. Store at −80 °C for up to 40 days.

2. Transformation (*see* **Note 13**). Defrost the competent cells on ice (10–20 min) and add 5 μL (*see* **Note 21**) of the phospho-ligation reaction mix. Swirl the tubes gently and incubate on ice for 30 min. Swirl the tubes gently and heat-shock cells in a water bath at 42 °C for 45 s and immediately transfer to ice for 10–15 min. Add 800 μL of fresh LB medium and incubate for 1 h at 37 °C, 200 rpm. Centrifuge for 3 min at 3000 × g, room temperature, remove 850 μL of the supernatant and gently resuspend the pellet in the remaining solution. Spread-plate the remaining ~150 μL solution on LB + ampicillin agar plates and incubate overnight at 37 °C. A positive transformation control with 1 μL of 1 pg/μL pUC18 plasmid and a negative process control with 1 μL of the PCR negative control should also be carried out.

3. Select three transformant colonies and inoculate into 5 mL LB + ampicillin medium (*see* **Note 14**) in a 15 mL Falcon tube. Incubate at 37 °C, 200 rpm overnight. No colonies should be visible for the negative process control. The LB + ampicillin plate for the positive transformation control should have approximately 50 colonies.

4. Isolate plasmid from cultures with a commercial plasmid purification kit and forward for sequencing of the insert in both directions. Greater than 80% of the sequences should contain the desired mutation and no other undesired mutation (*see* **Note 22**).

4   Notes

1. It is essential that the template used for iPCR is purified, circular, double-stranded DNA isolated from a *dam⁺ E. coli* strain. The majority of commonly used *E. coli* strains are *dam⁺*, including *E. coli* XL1-Blue, DH5α and JM109. *E. coli* JM110 and SCS110 are examples of dam⁻ strains and should not be used. *dam⁺ E. coli* strains contain the enzyme Dam methylase which methylates adenine residues in the sequence GATC. This methylated sequence is the target for digestion by *Dpn*I and allows for later removal of template DNA from the nonmethylated in vitro produced iPCR product.

   While best results are achieved with smaller templates, iPCR of templates up to 10 kb is commonplace, with some reports of successes with even larger plasmid constructs.

2. We have successfully used desalted primers for SDM but did encounter an increased number of incomplete product sequences with missing nucleotides at the ligation site. To enhance the yield of full length sequences, HPLC or PAGE purified sequences are recommended. The former augments the content of full length primers (≥85% are full length) while the latter, PAGE, is more apt for ensuring the full length (≥90% are full length) of longer primers (>40 bp).

   Primers are frequently resuspended in water or Tris buffer supplemented with EDTA. Nonsupplemented Tris buffer is preferred as the pH of water is often slightly acidic and can lead to depurination while EDTA can interfere with downstream processes by sequestering essential cations. Do not store oligonucleotides in water at 4 °C. Prepare aliquots of 20 μM working stock, store at −20 °C and avoid repeated freezing and thawing.

   Phosphorylation of the iPCR product with polynucleotide kinase (*see* Subheading 3.2, **step 2**, phospho-ligation) eliminates the need for phosphorylated primers. Nevertheless, if preferred, these may be utilized and the polynucleotide kinase treatment omitted by substitution of this enzyme with 1 μL water during the phospho-ligation step.

3. While the polymerase used here is Phusion High Fidelity DNA Polymerase, any high-fidelity DNA polymerase with a high extension rate and proofreading activity ($3' \rightarrow 5'$ exonuclease activity) may be used, e.g., Q5-High Fidelity DNA polymerase, *Pfu* Turbo DNA polymerase, and KOD DNA polymerase. Nevertheless, one should be aware of the particular template size limitations of the polymerase chosen, e.g., KOD DNA polymerase is recommended for templates ≤6 kbp; *Pfu* Turbo and Phusion DNA polymerases are reported to be able to amplify plasmids up to 15 kb. For best PCR results, use the

hot start variants of these polymerases where incorporation of automatic hot start technology permits polymerase activity at high temperatures only. This minimizes nonspecific amplification and primer dimer formation at low temperatures during reaction setup and during the initial PCR cycle, and allows for room temperature reaction setup. Currently commercialized examples include Platinum SuperFi DNA Polymerase, Phusion Hot Start High-Fidelity DNA Polymerase, Q5 Hot Start High-Fidelity DNA Polymerase. In all cases, modify the protocol given in this chapter according to the manufacturers' recommendations for the particular polymerase used.

4. Two buffers are provided with Phusion Polymerase, a HF and a GC Buffer. The former is used as the default buffer for high-fidelity amplification as the error rate with this is lower than with the latter. However, GC Buffer can improve the performance with certain difficult or long templates, such as GC-rich templates or templates with complex secondary structures. For amplification of GC rich templates, use of 3% DMSO with HF buffer should be initially investigated. The GC buffer should only be used when HF buffer gives unsatisfactory results.

5. Autoclave ultrapure water to ensure sterility and inactivate residual nucleases (DNase).

6. The most popular buffers for DNA electrophoresis are TAE and TBE (1 M Tris base, 1 M boric acid, and 0.02 M EDTA). Either may be used here and should give similar results.

7. Bromophenol blue is used as a tracking dye and has an approximate position on a 1% agarose gel equivalent to a 370 bp (TAE buffer) or 220 bp (TBE buffer) fragment. Xylene cyanol FF (0.03% in 6× loading buffer stock) may also be used and has an approximate position on a 1% agarose gel equivalent to a 4160 bp (TAE buffer) or 3030 bp (TBE buffer) fragment.

8. Use a molecular weight marker with component DNA of sizes similar to the expected iPCR product size. We commonly use the GeneRuler 1 kb DNA Ladder.

9. The restriction enzyme *Dpn*I digests template DNA (from *dam*+ strains) at the methylated sequence $G^{m6}ATC$, thereby "enriching for" the nonmethylated in vitro amplified iPCR product. *Dpn*I is highly active in the majority of commonly used polymerase reaction buffers (Phusion, Q5, etc.) and therefore the digestion can be performed directly in the PCR mix without any purification of the DNA. Recently, an optimized, three enzyme mix (*Dpn*I, polynucleotide kinase and ligase) has been reported for a more rapid (5 min) enrichment and phospho-ligation of iPCR products in one step (New England Biolabs).

10. ATP, DTT, and Mg^{2+} are essential buffer components for phospho-ligation. We commonly use T4 DNA ligase buffer

for the double T4 polynucleotide kinase-T4 DNA ligase reaction. Other buffers, such as the T4 polynucleotide kinase buffer, or FastDigest Buffer, or even many of the standard low salt restriction enzyme buffers supplemented with 1 mM riboATP may also be used. Oxidized DTT leads to reduced enzyme activity, avoid repeated freezing and thawing and avoid using solutions more than 1 year old. Addition of 5% polyethylene glycol (PEG) may enhance phospho-ligation but in this case extended ligation should be avoided.

11. T4 polynucleotide kinase phosphorylates the 5′-hydroxyl terminus of double and single stranded DNA and RNA. It is inhibited by ammonium ions and by high salt and high phosphate concentrations, do not use DNA precipitated with ammonium ions.

12. T4 DNA ligases join the 5′ phosphorylated and 3′ hydroxyl ends of the linear product to give a recircularized iPCR product. It is sensitive to high salt and high EDTA concentrations. Rapid ligases (Quick Ligase) which are reported to enable reaction completion in as little as 5 min have recently been marketed.

13. We commonly use in-house prepared chemically competent *E. coli* XL1-Blue as the cloning host. The preparation procedure described here allows for transformation efficiencies of $10^7 - 10^8$ cfu/μL which is normally sufficient for our SDM protocol. *E. coli* XL1-Blue is resistant to tetracycline and hence is not suited for plasmids with tetracycline resistance markers. Other, commercial, higher-efficiency cloning hosts and supercompetent cells for a higher number of transformants may also be used. In addition, use of electrocompetent hosts for transformation by electroporation allows for higher transformation efficiencies and is especially suited for large plasmids (~10 kbp and higher). In this latter case it is essential that the phospho-ligated circular DNA sample is purified (e.g., with a commercial DNA purification kit) to remove salts etc. prior to electroporation.

14. Ensure that the antibiotic/selection agent used is appropriate for the selective marker of the plasmid.

15. The description given is for a point mutation, but a similar primer design strategy may be used for short (1–3 bp) multiple base pair mutations or insertions, which may be included on one or both primers (*see* Fig. 2). Large insertions may be made by adding the nucleotides to be inserted on the 5′ ends of one, or both, of the inverse primers (*see* Fig. 2). Here, the perfectly matched portion of the primers should be 24–30 bp in length and should be used for calculation of the primer melting temperature. For deletions, the inverse primers should be designed to be perfectly matched to the sequences flanking

the fragment to be deleted (*see* Fig. 2). All remaining steps of the iPCR SDM procedure for these different types of mutations are similar to that described.

16. Normal considerations for PCR primer design should be adhered to, i.e., forward and reverse primers should have similar (<5 °C difference) melting temperatures (*see* **Note 11**), a GC content of 40–60%, 1 or 2 Gs or Cs at the 3′ end, and direct repeats, secondary structures, primer dimers and mispriming should be avoided. When designing mutations for introduction of an amino-acid change, the codon usage of the expression host should be taken into consideration for selection of the codon most favoured by the host or/and which requires the least number of base changes. Primer design programs are recommended, e.g., Primer3Plus, OligoCalc, SnapGene, NEBaseChanger, and OligoPerfect.

17. The optimum reaction conditions vary considerably with the polymerase and buffer system used, therefore the reaction conditions recommended by the supplier of the chosen polymerase should always be employed. Mainly, this involves alterations in the amount of polymerase and buffer used.

18. The optimum PCR cycle conditions vary with the polymerase and buffer system used, e.g., the phusion DNA polymerase system is characterized by elevated denaturation and annealing temperatures and high extension rate as compared to the majority of other polymerases. Therefore, the reaction temperatures and times recommended by the manufacturers of the chosen polymerase should always be employed.

    Usually, high-fidelity polymerases are thermostable at temperatures higher than 98 °C. Therefore, denaturation temperatures from 95 to 98 °C can be used. The shortest denaturation time should be used so as to avoid template damage. For most templates a 30-s initial denaturation from 95 to 98 °C is enough. Some templates, due to higher complexity, may require up to 3 min, or up to 5 min for GC-rich templates (>70% GC content).

    The most appropriate annealing temperature varies widely with the polymerase system employed and should be calculated as recommended by the supplier. Free online calculators for determination of the annealing temperatures are provided for the various DNA polymerase systems being currently commercialized, e.g., the Phusion Tm Calculator at ThermoFisher Scientific. For primers with calculated annealing temperatures ≥72 °C with Phusion, a two-step thermocycling protocol is recommended in which the annealing step is eliminated.

    The extension time and temperature depends on the extension rate and optimal temperature of the polymerase utilized, as well as the amplicon length and complexity. Most commonly, 72 °C is used. The extension time employed should ensure adequate full-length product synthesis and 15–60 s/kb is usually sufficient.

We recommend using 25 cycles. A higher number of cycles (up to 35) may increase product yield but also increases the probability of secondary, unwanted mutations.

19. If, in addition to a DNA band of the desired size, other nonspecific DNA bands are visible for the sample, then all the remaining 15 μL of the reaction should be run on a 1% agarose gel and the desired DNA band size excised and purified with a commercial gel extraction DNA purification kit. The purified DNA fragment can then be used directly in the step "Template Removal and Product Recircularization" (Subheading 3.2).

The absence of any visible bands indicates PCR failure and hence typical PCR troubleshooting procedures should be followed, e.g., check primers design, reduce the annealing temperature by 3–5 °C increments, optimize Mg^{2+} concentration in 0.5 mM increments, increase denaturation and extension times, repeat experiment with various concentrations of template. In the case of phusion polymerase, use of both HF and GC buffers as well as addition of 3% DMSO should first be investigated (*see* **Note 4**).

A weak band may be visible for the negative control if higher template concentrations were used (≥ 10 ng) but this should be many fold weaker than the sample band.

20. Avoid using large volumes of PCR product as this may interfere with the subsequent phospho-ligation and transformation. If a poor PCR yield leads to the need for larger volumes of PCR product, this should first be purified using a commercially available DNA purification kit.

Improved phospho-ligation may be attained by use of 5% PEG. Also, following polynucleotide kinase addition, the sample may be incubated at 37 °C for 30 min before cooling to room temperature, adding the T4 DNA ligase, and further incubating at room temperature for 90 min.

21. Avoid using phospho-ligation mix volumes that are more than 10% of the competent cell volume as this leads to a reduced transformation efficiency. Purify phospho-ligation mix by use of a commercial DNA purification kit if transforming by electroporation.

22. The insert sequence should contain the desired mutation only. While it is not feasible to sequence the entire plasmid construct, the use of a high fidelity polymerase reduces the risk of secondary mutations in the vector sequence. To ensure the absence of such mutations, the insert sequence may be recloned into the original non-PCR-amplified vector.

References

1. Bommarius AS, Paye MF (2013) Stabilizing biocatalysts. Chem Soc Rev 42:6534–6565. doi:10.1039/c3cs60137d

2. Collins T, De Vos D, Hoyoux A, Savvides SN, Gerday C, Van Beeumen J, Feller G (2005) Study of the active site residues of a glycoside hydrolase family 8 xylanase. J Mol Biol 354(2):425–435

3. Davids T, Schmidt M, Bottcher D, Bornscheuer UT (2013) Strategies for the discovery and engineering of enzymes for biocatalysis. Curr Opin Chem Biol 17(2):215–220. doi:10.1016/j.cbpa.2013.02.022

4. Sanjuan R (2010) Mutational fitness effects in RNA and single-stranded DNA viruses: common patterns revealed by site-directed mutagenesis studies. Philos Trans R Soc Lond B Biol Sci 365(1548):1975–1982. doi:10.1098/rstb.2010.0063

5. Steiner K, Malke H (1997) Primary structure requirements for in vivo activity and bidirectional function of the transcription terminator shared by the oppositely oriented skc/rel-orf1 genes of *Streptococcus equisimilis* H46A. Mol Gen Genet 255(6):611–618

6. Traboni C, Ciliberto G, Cortese R (1982) A novel method for site-directed mutagenesis: its application to an eukaryotic tRNAPro gene promoter. EMBO J 1(4):415–420

7. Hutchison CA, Phillips S, Edgell MH, Gillam S, Jahnke P, Smith M (1978) Mutagenesis at a specific position in a DNA sequence. J Biol Chem 253(18):6551–6560

8. Hutchison CA 3rd, Edgell MH (1971) Genetic assay for small fragments of bacteriophage phi X174 deoxyribonucleic acid. J Virol 8(2):181–189

9. Edgell MH, Hutchison CA 3rd, Sclair M (1972) Specific endonuclease R fragments of bacteriophage phiX174 deoxyribonucleic acid. J Virol 9(4):574–582

10. Reeves AR (ed) (2016) In vitro mutagenesis. Methods and protocols, series vol 1498, 1 edn. Humana, New York, NY. doi:10.1007/978-1-4939-6472-7

11. Ho SN, Hunt HD, Horton RM, Pullen JK, Pease LR (1989) Site-directed mutagenesis by overlap extension using the polymerase chain reaction. Gene 77(1):51–59

12. Stuckey S, Storici F (2013) Gene knockouts, in vivo site-directed mutagenesis and other modifications using the delitto perfetto system in *Saccharomyces cerevisiae*. Methods Enzymol 533:103–131. doi:10.1016/B978-0-12-420067-8.00008-8

13. Kren BT, Bandyopadhyay P, Steer CJ (1998) In vivo site-directed mutagenesis of the factor IX gene by chimeric RNA/DNA oligonucleotides. Nat Med 4(3):285–290

14. Higuchi R, Krummel B, Saiki RK (1988) A general method of in vitro preparation and specific mutagenesis of DNA fragments: study of protein and DNA interactions. Nucleic Acids Res 16(15):7351–7367

15. Triglia T, Peterson MG, Kemp DJ (1988) A procedure for in vitro amplification of DNA segments that lie outside the boundaries of known sequences. Nucleic Acids Res 16(16):8186

16. Ochman H, Gerber AS, Hartl DL (1988) Genetic applications of an inverse polymerase chain reaction. Genetics 120(3):621–623

17. Hemsley A, Arnheim N, Toney MD, Cortopassi G, Galas DJ (1989) A simple method for site-directed mutagenesis using the polymerase chain reaction. Nucleic Acids Res 17(16):6545–6551

18. Qi D, Scholthof KB (2008) A one-step PCR-based method for rapid and efficient site-directed fragment deletion, insertion, and substitution mutagenesis. J Virol Methods 149(1):85–90. doi:10.1016/j.jviromet.2008.01.002

19. Zheng L, Baumann U, Reymond JL (2004) An efficient one-step site-directed and site-saturation mutagenesis protocol. Nucleic Acids Res 32(14):e115. doi:10.1093/nar/gnh110

20. Liu H, Naismith JH (2008) An efficient one-step site-directed deletion, insertion, single and multiple-site plasmid mutagenesis protocol. BMC Biotechnol 8:91. doi:10.1186/1472-6750-8-91

21. Xia Y, Chu W, Qi Q, Xun L (2015) New insights into the QuikChange process guide the use of Phusion DNA polymerase for site-directed mutagenesis. Nucleic Acids Res 43(2):e12. doi:10.1093/nar/gku1189

Chapter 6

Synthesis of Fusion Genes for Cloning by Megaprimer-Based PCR

Tatiana Q. Aguiar, Carla Oliveira, and Lucília Domingues

Abstract

The polymerase chain reaction (PCR) is the technique of choice used to obtain DNA for cloning, because it rapidly provides high amounts of desired DNA fragments and allows the easy introduction of extremities adequate for enzyme restriction or homologous recombination, and of artificial, native, or modified sequence elements for specific applications. In this context, the use of megaprimer-based PCR strategies allows the versatile and fast assembly and amplification of tailor-made DNA sequences readily available for cloning.

In this chapter, we describe the design and use of a megaprimer-based PCR protocol to construct customized fusion genes ready for cloning into commercial expression plasmids by restriction digestion and ligation.

Key words Megaprimers, Two-step PCR, Fusion genes, Molecular cloning, Restriction enzymes, Expression plasmids

1 Introduction

The polymerase chain reaction (PCR) is one of the most important basic methodologies used to obtain DNA fragments for cloning [1]. Typically, the DNA fragment/gene of interest has to be amplified from genomic or vector DNA by PCR before it can be cloned into a cloning/expression vector. The first step in this procedure is the design of the necessary primers. These must contain 17–30 base pair (bp)-long sequences complementary to the fragment/gene of interest at their 3′-end and can additionally have one or several of the following elements, noncomplementary to the template DNA, in their 5′ region: (a) restriction(s) site(s) for cloning into plasmid vectors [1–7]; (b) some nucleotides (typically 2–10) upstream of the restriction site, necessary to increase the cleavage efficiency of enzymes that generally have low cleavage efficiency close to the end of a DNA fragment [1–7]; (c) start, stop, or modified codons [1–3]; (d) DNA sequences

Lucília Domingues (ed.), *PCR: Methods and Protocols*, Methods in Molecular Biology, vol. 1620,
DOI 10.1007/978-1-4939-7060-5_6, © Springer Science+Business Media LLC 2017

encoding linkers, specific chemical or enzymatic cleavage sites, fusion tags and/or peptides/proteins [1–6]; (e) overlap sequences to DNA regions of interest (typically 20–40 bp-long), necessary in PCR-based DNA assembly and cloning methods [1], and for the construction of PCR-based genomic integration/disruption cassettes [7]; etc.

Primers that contain long extra 5′ sequences are generally larger (typically up to 300 bp) than conventional primers (17–40 bp) and, therefore, are called megaprimers. Megaprimers can be used to define the 5′-, 3′-end, or both ends of a PCR product. In a PCR reaction, megaprimers can be used together or coupled with conventional primers, as long as the difference between the melting temperature (T_m) of the complementary region of the two primers is within 5 °C (*see* [1, 8] for basic rules in primer design).

Traditionally, megaprimers have been obtained using PCR-based strategies [8, 9], but nowadays they can also be obtained by chemical synthesis at an affordable price. Megaprimers have been mainly used for PCR-based site-directed mutagenesis [8, 9], but they also have applicability in the construction of gene fusions [3–5], as well as in the synthesis of short DNA sequences [3–5].

In this chapter, we describe a megaprimer-based PCR protocol to construct gene fusions for subsequent cloning by restriction digestion and ligation into commercially available expression plasmids. The presented case study refers to the fusion and amplification of the complete gene sequence of the carbohydrate-binging module 1 (CBM1) from *Trichoderma reesei* Cel7A (Cellobiohydrolase I), with (180 bp) and without (108 bp) its native linker (NL), to the gene sequence of the reporter Emerald Green Fluorescent Protein (EmGFP; 717 bp) by PCR, using manually designed megaprimers. The *EmGFP-TrCBM1*$_{Cel7A}$ fusion genes, with and without the CBM1 NL, were assembled and amplified by two-step primer extension PCR with megaprimers containing partial CBM1-encoding sequences with codons optimized for use in the expression host *Pichia pastoris* (Table 1), using as initial template the EmGFP-coding gene from plasmid pPCG [10]. In the designed megaprimers, a sequence encoding a recognition site for the Tobacco Etch Virus (TEV; 21 bp) protease was included between the genes, to allow the obtainment of free EmGFP and CBM1 when necessary, and a histidine tag (18 bp) was fused to the C-terminal of the CBM1 sequence, to allow subsequent purification by Immobilized Metal Affinity Chromatography (IMAC). Other sequence elements of interest, such as restriction sites and stop codon, were also added to the primers. The strategy used is schematized in Fig. 1. As described herein, this strategy is adequate to synthesize and fuse genes with up to 250 bp to a given template gene, but it may be easily adapted to synthesize and fuse motifs of any molecular weight to any template gene, by using as many megaprimers and PCR rounds as necessary.

Table 1
Primers used in this work. The melting temperature (T_m) indicated for each primer is the T_m of the sequence complementary to the corresponding template DNA

	Primer	Sequence (5′ → 3′)[a]	Length (bp)	T_m (°C)
P_1	EmGFP_FW	*cg*gaattcatggtgagcaagggcgag	26	59
P_2	*Tr* CBM1$_{Cel7A}$_RV1.1	CCACCACATTGACCGTAATGAGATTGAGTACCTTGAAA GTACAAGTTTTCcttgtacagctcgtccatgc	70	56
P_3	*Tr* CBM1$_{Cel7A}$_RV1.2	CCACCACATTGACCGTAATGAGATTGAGTTGGACCTG **GAGAAGAACCAGTAGTAGTAGCTGGTCTTCTA GTAGTAGTAGTACCTCTGTTACCACCTGGTGG** ACCTTGAAAGTACAAGTTTTCcttgtacagctcgtccatgc	142	56
P_4	*Tr* CBM1$_{Cel7A}$_RV2	*gtgggtacc**TCA*** *atgatgatgatgatgatgc*CAAACATTGAGAGTAGTA TGGGTTCAAAACTTGACAAGTAGTACCAGAAGCA CAAACAGTTGGACCAGAGTAACCAATAccaccacattgaccgtaatgaga	131	57

[a]*Simple lower* and *upper case* refer to the sequences complementary to the EmGFP- and CBM1-coding regions, respectively. *Bold upper case* refers to the sequence encoding the CBM1 native linker. *Underlined upper* and *lower case* refer to the sequence encoding a TEV recognition site and a six histidines tag, respectively. The stop codon is indicated in *upper case italics*. The recognition sites for the selected restriction enzymes are indicated in *lower case bold* (*EcoRI*: **gaattc** and *KpnI*: **ggtacc**) and the nucleotides added upstream of them are in *lower case italics*

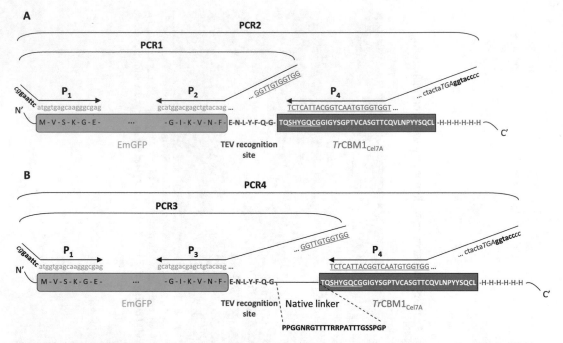

Fig. 1 Illustration of the megaprimer-based PCR strategy used to assemble the fusion genes *EmGFP-TrCBM1*$_{Cel7A}$ (**a**) and *EmGFP-NL-TrCBM1*$_{Cel7A}$ (**b**), with indication of the location of primers and of all the extra sequence elements added after each PCR. Legends for the represented primer sequences: *black lower case*, recognition sites for selected restriction enzymes (*Eco*RI: gaattc and *Kpn*I: ggtacc); *green lower case*, sequences complementary to the EmGFP-coding region; *golden, underlined upper case*, complementary region of primer P$_4$ with PCR1/PCR3 products; *blue lower case*, histidine codons; *red upper case italics*, stop codon (TGA)

---

2   Materials

2.1   Equipment

1. Thermal cycler.
2. Agarose gel electrophoresis apparatus.
3. Microwave.
4. UV transiluminator with a gel documentation unit.
5. Microvolume UV–Vis spectrophotometer for nucleic acid quantification.
6. Tabletop centrifuge.
7. Pipettes and pipette tips.
8. Microtubes (0.2 mL PCR tubes and 1.5 mL microcentrifuge tubes).
9. Temperature-controlled water bath.

2.2   Supplies and Reagents

1. Sterile ultrapure water.
2. Template DNA: Plasmid DNA (from pPCG [10]) isolated using GenElute™ Plasmid Miniprep Kit (Sigma-Aldrich), or equivalent, and purified PCR products.

3. Oligonucleotide primers: Ordered from a local supplier at a 20 nmol synthesis scale, and without HPLC or gel purification. Aliquots of 25 µM working solution were prepared from 100 µM stock solution and stored at −20 °C to prevent contamination of stock and repeated freeze–thaw cycles.

4. Proofreading DNA polymerase: In this protocol we used Vent® DNA polymerase, supplied with 10× ThermoPol® Reaction Buffer and 100 mM MgSO$_4$ solution (New England Biolabs).

5. dNTPs, 10 mM each.

6. 1× Tris–Acetate–EDTA (TAE) buffer: 40 mM Tris, 20 mM acetic acid, 1 mM EDTA, pH 8.3.

7. 1% (w/v) agarose gels (in 1× TAE buffer) with 0.05% (v/v) GreenSafe Premium (NZYTech), or equivalent.

8. 10 kb DNA ladder.

9. 6× DNA loading buffer: 30% (w/v) glycerol, 0.25% (w/v) bromophenol blue.

10. Commercial kits for DNA purification: QIAquick® Gel Extraction and PCR Purification Kit (Qiagen), or equivalent.

11. Vector cloning: Plasmid pPICZαA (Invitrogen); restriction enzymes *Eco*RI-HF, *Kpn*I-HF and 10× CutSmart® buffer (New England Biolabs), or equivalent; T4 DNA ligase and 10× ligation buffer; *Escherichia coli* chemically competent cloning cells and SOC medium; LB agar plates containing appropriated antibiotics.

3 Methods

3.1 Primer Design

Rational primer design is a critical step for the success of this protocol, which requires the prior knowledge of the DNA sequences of the genes and elements that we desire to fuse, as well as the maps of the expression vectors where we intend to clone the constructed fusions.

1. Assemble the final base pair sequence of the desired fusion genes and optimize their codons according to the preferential codon usage of the host where they are going to be expressed in (Fig. 2). In this protocol, the EmGFP-coding gene from pPCG [10] was chosen as template for the initial PCR and the codons of the sequences that were fused to this gene were optimized for expression in *P. pastoris* using the online software GENEius (Eurofins Genomics) (*see* **Note 1**).

2. Select appropriate enzymes for cloning. Analyse the multiple cloning site of the selected expression vector and the restriction map of each fusion gene (e.g., with the online NEBcutter tool, New England Biolabs). Select two enzymes that do not

(A) *EmGFP-TrCBM1$_{Cel7A}$*

```
atggtgagcaagggcgaggagctgttcaccggggtggtgcccatcctggtcgagctggacggcgacgtaaacggccaca
agttcagcgtgtccggcgagggcgagggcgatgccacctacggcaagctgaccctgaagttcatctgcaccaccggcaa
gctgcccgtgccctggcccaccctcgtgaccaccttgacctacggcgtgcagtgcttcgcccgctaccccgaccacatg
aagcagcacgacttcttcaagtccgccatgcccgaaggctacgtccaggagcgcaccatcttcttcaaggacgacggca
actacaagacccgcgccgaggtgaagttcgagggcgacaccctggtgaaccgcatcgagctgaagggcatcgacttcaa
ggaggacggcaacatcctggggcacaagctggagtacaactacaacagccacaaggtctatatcaccgccgacaagcag
aagaacggcatcaaggtgaacttcaagacccgccacaacatcgaggacggcagcgtgcagctcgccgaccactaccagc
agaacacccccatcggcgacggccccgtgctgctgcccgacaaccactacctgagcacccagtccgccctgagcaaaga
ccccaacgagaagcgcgatcacatggtcctgctggagttcgtgaccgccgccgggatcactctcggcatggacgagctg
tacaagGAAAACTTGTACTTTCAAGGTACTCAATCTCATTACGGTCAATGTGGTGGTATTGGTTACTCTGGTCCAACTG
TTTGTGCTTCTGGTACTACTTGTCAAGTTTTGAACCCATACTACTCTCAATGTTTGcatcatcatcatcatcatTGA
```

(B) *EmGFP-NL-TrCBM1$_{Cel7A}$*

```
atggtgagcaagggcgaggagctgttcaccggggtggtgcccatcctggtcgagctggacggcgacgtaaacggccaca
agttcagcgtgtccggcgagggcgagggcgatgccacctacggcaagctgaccctgaagttcatctgcaccaccggcaa
gctgcccgtgccctggcccaccctcgtgaccaccttgacctacggcgtgcagtgcttcgcccgctaccccgaccacatg
aagcagcacgacttcttcaagtccgccatgcccgaaggctacgtccaggagcgcaccatcttcttcaaggacgacggca
actacaagacccgcgccgaggtgaagttcgagggcgacaccctggtgaaccgcatcgagctgaagggcatcgacttcaa
ggaggacggcaacatcctggggcacaagctggagtacaactacaacagccacaaggtctatatcaccgccgacaagcag
aagaacggcatcaaggtgaacttcaagacccgccacaacatcgaggacggcagcgtgcagctcgccgaccactaccagc
agaacacccccatcggcgacggccccgtgctgctgcccgacaaccactacctgagcacccagtccgccctgagcaaaga
ccccaacgagaagcgcgatcacatggtcctgctggagttcgtgaccgccgccgggatcactctcggcatggacgagctg
tacaagGAAAACTTGTACTTTCAAGGTCCACCAGGTGGTAACAGAGGTACTACTACTACTAGAAGACCAGCTACTACTA
CTGGTTCTTCTCCAGGTCCAACTCAATCTCATTACGGTCAATGTGGTGGTATTGGTTACTCTGGTCCAACTGTTTGTGC
TTCTGGTACTACTTGTCAAGTTTTGAACCCATACTACTCTCAATGTTTGcatcatcatcatcatcatTGA
```

Fig. 2 DNA sequences of the customized fusion genes *EmGFP-TrCBM1$_{Cel7A}$* (**a**) and *EmGFP-NL-TrCBM1$_{Cel7A}$* (**b**) with the codons optimized for expression in *P. pastoris*. The EmGFP- and CBM1-coding regions are highlighted in *green* and *gold*, respectively. The CBM1 NL-coding region is highlighted in *grey*. The sequences encoding a TEV recognition site and a six histidines tag are highlighted in *pink* and *blue*, respectively. Highlighted in *red* is the stop codon. The *underlined regions* indicate the priming sites for the designed primers

cut in the fusion genes and allow their directional cloning into the desired vector region in frame with the existing reading frame (*see* **Note 2**). In this protocol, the enzymes *Eco*RI and *Kpn*I were selected to directionally clone the *EmGFP-TrCBM1$_{Cel7A}$* and *EmGFP-NL-TrCBM1$_{Cel7A}$* fusions in frame with the *Saccharomyces cerevisiae* α-factor signal peptide-coding sequence in the commercial expression plasmid pPICZαA (Invitrogen).

3. After analysing the sequences of the customized fusion genes, select appropriate priming regions (Fig. 2). For this, standard primer design rules should be applied [1, 8], including: minimum of 17 bp complementary to template DNA, GC content between 40 and 60%, T_m preferably between 52 and 62 °C, T_m difference between primer pairs <5 °C, presence of 1–3 G's or C's within the last five bases from the 3′-end of primers, avoid potential primer dimmer (ΔG > −7.5 kcal/mol) and hairpin

formation ($\Delta G > -3$ kcal/mol) as well as false priming sites. Use computational tools to help you estimate the T_m of the selected priming regions (for Vent® DNA polymerase use the online NEB T_m calculator, New England Biolabs) and change their length (always respecting the standard primer design rules) until you reach a satisfactory T_m. Note that this estimated T_m is only for the primer region complementary to the template DNA and not for the full-length primer.

4. For the full-length primers, check for GC content, potential primer dimer and hairpin formation (e.g., with the online OligoAnalyzer tool, Integrated DNA Technologies). If any of these parameters are outside of the recommended guidelines, proceed to the necessary adjustments (*see* **Note 3**).

5. In this protocol, the EmGFP-coding gene served as initial template to assemble each of the fusion genes. Thus, we designed one conventional forward primer (P_1), complementary to the 5′-end region of the EmGFP-encoding gene, which could be used in combination with different reverse megaprimers (Table 1, Fig. 1). The reverse megaprimers P_2 and P_3 were designed to anneal at the same priming site in the 3′-end region of the EmGFP-encoding gene and to have equal 5′-end sequences that were used as priming site for primer P_4 (Table 1, Figs. 1 and 2). However, while primer P_3 was designed to include a coding sequence for a TEV recognition site plus the CBM1 NL between its 3′- and 5′-ends, in primer P_2 only the first sequence was included (Table 1, Figs. 1 and 2). The megaprimer P_4 was designed to contain a 3′-end sequence complementary to the 3′ region of the PCR products generated with primes P_1 and P_2/P_3 and to include in its 5′ region the remaining coding sequence for CBM1 plus six histidine codons and a stop codon (*see* **Note 4**). Restriction sites for *Eco*RI and *Kpn*I were added to the 5′-end of primers P_1 and P_4, respectively, preceded by two nucleotides to increase the cleavage efficiency of the final PCR products by these enzymes.

3.2 Double-Step Primer Extension PCR

A two-step PCR was performed to synthesize the fusion genes *EmGFP-TrCBM1*$_{Cel7A}$ and *EmGFP-NL-TrCBM1*$_{Cel7A}$, in which the PCR products from the first PCR round served as template for the second PCR (Table 2).

1. To obtain the PCR1 and PCR3 products (Table 2, Fig. 1), mix the following components in PCR tubes:
   - Template DNA – 100–300 ng (in 1 μL).
   - Primer P_1 (25 μM) – 1 μL.
   - Primer P_2 for PCR1 and P_3 for PCR3 (25 μM) – 1 μL.
   - dNTPs (10 mM each) – 1.5 μL.

Table 2
Primers, template DNA, and annealing temperature (T_a) used for the assembly and amplification of the fusion genes *EmGFP-TrCBM1$_{Cel7A}$* and *EmGFP-NL-TrCBM1$_{Cel7A}$*

Fusion gene	PCR products	Primers	Template	T_a (°C)
EmGFP-TrCBM1$_{Cel7A}$	PCR1	P_1 P_2	pPCG[a]	55
	PCR2	P_1 P_4	PCR1 product	53
EmGFP-NL-TrCBM1$_{Cel7A}$	PCR3	P_1 P_3	pPCG[a]	50
	PCR4	P_1 P_4	PCR3 product	53

[a]Plasmid harboring the EmGFP-coding gene [10]

- 10× ThermoPol® Reaction Buffer – 5 μL.
- 100 mM $MgSO_4$ – 1 μL.
- Ultrapure H_2O – 39 μL.
- Vent® DNA polymerase – 0.5 μL.

2. Step up the thermal cycler using the following parameters:

Number of cycles	PCR step	Temperature (°C)	Time
1	Initial denaturation	95	2 min
30	Denaturation	95	2 min
	Annealing	(*see* Table 2)	45 s
	Extension	72	1 min/kb
1	Final extension	72	10 min

3. After the amplification, assess the DNA fragments' size and quality by running 5 μL of each PCR sample on 1% (w/v) agarose gel. If the expected PCR products are not observed refer to **Note 5**. If the PCR products are the right size, but a primer dimer band is also present (typically, below 200 bp), or unspecific PCR products are formed, perform agarose gel purification of the PCR products with the QIAquick® Gel Extraction kit (spin protocol) before proceeding to the next step (*see* **Note 6**). If the PCR products are the right size and are present as a single band perform silica-column purification of the products with the QIAquick® PCR Purification Kit (*see* **Note 6**). Determine the DNA concentration.

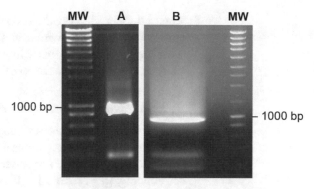

Fig. 3 Final PCR products from the amplification of the fusion genes *EmGFP-TrCBM1$_{Cel7}$* (**A**, PCR2 product) and *EmGFP-NL-TrCBM1$_{Cel7A}$* (**B**, PCR4 product). *MW* molecular weight standards

4. To obtain the PCR2 and PCR4 products, mix the components described in **step 1**, using the same amounts, but replace primers P$_2$ and P$_3$ with primer P$_4$, and use as template DNA the adequate purified PCR product from **step 3** (Table 2). Since the DNA purification procedure reduces the PCR product yield, a high volume of PCR product (up to 15 µL) is generally needed as template DNA (*see* **Note 7**).

5. Purify the PCR products as described in **step 3** (Fig. 3) (*see* **Note 8**).

3.3 Vector Cloning and Confirmation of the Fusion Genes' Sequence

The final PCR products can be digested with the cloning enzymes and cloned directly into the expression vector or can alternatively be cloned into an intermediate cloning vector, propagated, excised by restriction digestion, and finally cloned into the expression vector. Many convenient commercial systems for cloning of PCR products are available, either for blunt-ended (e.g., pMOS®, GE healthcare) or A-ended products (e.g., pGEM®-T, Promega). Thermostable DNA polymerases with proofreading activity, such as Vent®, generate blunt-ended fragments, but these can be modified using an A-tailing procedure (*see* **Note 9**), and thus cloned in any type of PCR cloning vectors. However, direct cloning is a more rapid and less expensive approach.

1. Digest overnight at 37 °C the purified PCR2 and PCR4 products with the cloning enzymes *Eco*RI-HF and *Kpn*I-HF, using the following reaction mixture:
   - Purified PCR product – 25 µL
   - *Eco*RI-HF – 1 µL
   - *Kpn*I-HF – 1 µL
   - 10× CutSmart® buffer – 3 µL

2. Purify the digested products with the QIAquick® PCR Purification Kit (*see* **Note 6**).

3. Ligate the digested products to the pPICZαA expression plasmid (previously digested with the same enzymes and gel purified) with T4 DNA ligase, following the instructions of the manufacturer and using a molar ratio of 1:3–1:10 (*see* **Note 10**).

4. Transform *E. coli* chemically competent cells with the entire ligation mixture according to the manufacturer's instructions.

5. Screen for positive clones (e.g., by colony PCR), propagate cells from one of these clones and isolate plasmid DNA to subsequently confirm the sequence of the inserted fusion genes by sequencing (*see* **Note 11**). If the sequence and reading frame of the fusion genes are correct, the recombinant plasmids are ready to be transformed into the final expression host (in this case *P. pastoris*). If not, review your strategy and repeat this protocol from the step you consider most adequate.

4   Notes

1. There are several codon optimization tools available for adapting DNA sequences to the codon preference of a particular expression host (consult Table 4.3 in [1]). Codon usage tables for several host organisms are also available from public databases (e.g., Codon Usage Database).

2. When cloning genes into a commercial expression vector, special care must be taken to maintain the existing reading frame, so that the ultimately expressed protein has the desired sequence. This is particularly critical when the restriction enzymes used for cloning have within their recognition sequence a start codon, as is the case for *Nco*I (5′ C/CATGG 3′). The ATG within this site can be used directly to create the ATG start codon and/or the ATG codon for a methionine residue. However, this restriction site dictates that the first nucleotide of the next triplet codon must be a G, and that two extra bases are necessary to maintain the reading frame, which can be added in primer design. Thus, an extra amino acid will be added to the protein. One way to overcome this limitation is to use enzymes that, while recognizing different nucleotide sequences, are able to generate ends compatible with those generated by the *Nco*I enzyme (e.g., *Bsa*I).

3. The design of primers for gene cloning into expression plasmids has several constraints, such as limited flexibility in the location of priming regions (the primers must anneal at the 5′ and 3′ extremities of the coding region), and difficulty in fulfilling the requirements for secondary structures formation when restriction sites or other extra sequences are added to the 5′-end of primers. Therefore, several adjustments generally have to be manually performed using a trial and error method

until the most satisfactory parameters for the designed primers can be obtained, which many times are not the optimal ones.

4. We could have constructed longer megaprimers P_2 and P_3, instead of constructing megaprimer P_4, but it would have been cheaper to order the synthesis of the entire fusion genes than ordering the synthesis of oligos >150 bp. Moreover, the bigger the primer, the higher the probability for secondary structure formation, which has negative impact in PCR efficiency. Therefore, we chose to order shorter (and cheaper) megaprimers and perform a two-step PCR to fuse the desired genes.

5. Run a higher volume of the PCR sample (>5 µL) to verify if you can then observe the expected PCR product. Otherwise, it may be necessary to experimentally optimize the PCR conditions. As initial steps, a gradient of annealing temperatures and different concentrations of template DNA may be tested, as well as a superior number of PCR cycles (up to 40). Subsequently, different concentrations of primers and of $MgSO_4$ may also be tested.

6. In both DNA purification protocols, elute the products from the columns with the minimum amount of warm ultrapure water (typically 30 µL) and repeat the elution step with the 30 µL eluate to increase DNA recovery and maximize the concentration of the eluate.

7. The volume of the PCR product to use as template should be sufficiently high to allow amplification, but it should not inhibit the PCR reaction. We have obtained good results using template volumes ≤30% of the final reaction volume.

8. Although in Fig. 3 several unspecific products (such as primer dimers) are observed along with the expected PCR product, in this case it was not necessary to optimize the PCR conditions, because the band corresponding to the product with the desired molecular weight could be easily isolated from the others and the amount of the purified product was sufficient for cloning. However, to reduce primer dimer formation, lower primer concentration could have been used.

9. At the end of the PCR reaction, add to the reaction mixture 1 µL of Taq DNA polymerase and perform a 10-min extension step at 72 °C.

10. Since several rounds of purification reduce the amount of DNA available for the cloning procedure, sometimes it is necessary to use a final ligation reaction volume higher than the standard 10 µL indicated by the manufacturer. We have obtained good results with ligation reactions with up to 20 µL.

11. Alternatively, the sequence of the fusion genes can be confirmed by direct sequencing of the final PCR products.

However, when sequencing from the expression plasmid, it is also possible to confirm if the direction and reading frame of the sequences are correct.

Acknowledgments

This study was supported by the Portuguese Foundation for Science and Technology (FCT) under the scope of the strategic funding of UID/BIO/04469/2013 unit and COMPETE 2020 (POCI-01-0145-FEDER-006684), and BioTecNorte operation (NORTE-01-0145-FEDER-000004) funded by European Regional Development Fund under the scope of Norte2020– Programa Operacional Regional do Norte, Project RECI/BBB-EBI/0179/2012 (FCOMP-01-0124-FEDER-027462), Project GlycoCBMs PTDC/AGR-FOR/3090/2012 (FCOMP-01-0124-FEDER-027948), and grant SFRH/BDP/63831/2009 to Carla Oliveira.

References

1. Oliveira C, Aguiar TQ, Domingues L (2017) 4 – Principles of genetic engineering. In: Pandey A, Teixeira JA (eds) Current developments in biotechnology and bioengineering: Foundations of biotechnology and bioengineering, 1st edn. Elsevier, Oxford, pp 81–127

2. Oliveira C, Sepúlveda G, Aguiar TQ, Gama FM, Domingues L (2015) Modification of paper properties using carbohydrate-binding module 3 from the *Clostridium thermocellum* CipA scaffolding protein produced in *Pichia pastoris*: elucidation of the glycosylation effect. Cellulose 22:2755–2765

3. Guerreiro CI, Fontes CM, Gama M, Domingues L (2008) *Escherichia coli* Expression and purification of four antimicrobial peptides fused to a family 3 carbohydrate-binding module (CBM) from *Clostridium thermocellum*. Protein Expr Purif 59:161–168

4. Oliveira C, Felix W, Moreira RA, Teixeira JA, Domingues L (2008) Expression of frutalin, an α-D-galactose-binding jacalin-related lectin, in the yeast *Pichia pastoris*. Protein Expr Purif 60:188–193

5. Costa SJ, Almeida A, Castro A, Domingues L, Besir H (2013) The novel Fh8 and H fusion partners for soluble protein expression in *Escherichia coli*: a comparison with the traditional gene fusion technology. Appl Microbiol Biotechnol 97:6779–6791

6. Ramos R, Domingues L, Gama M (2010) *Escherichia coli* Expression and purification of LL37 fused to a family III carbohydrate-binding module from *Clostridium thermocellum*. Protein Expr Purif 7:1–7

7. Aguiar TQ, Dinis C, Domingues L (2014) Cre-loxP-based system for removal and reuse of selection markers in *Ashbya gossypii* targeted engineering. Fungal Genet Biol 68:1–8

8. Sambrook J, Russell DW (2001) Molecular cloning: a laboratory manual. Cold Spring Harbor Laboratory Press, Cold Spring Harbor, NY

9. Ling MM, Robinson BH (1997) Approaches to DNA mutagenesis: an overview. Anal Biochem 254:157–178

10. Wan W, Wang DM, Gao XL, Hong J (2011) Expression of family 3 cellulose-binding module (CBM3) as an affinity tag for recombinant proteins in yeast. Appl Microbiol Biotechnol 91:789–798

Chapter 7

A Novel Platform for High-Throughput Gene Synthesis to Maximize Recombinant Expression in *Escherichia coli*

Ana Filipa Sequeira, Joana L.A. Brás, Vânia O. Fernandes, Catarina I.P.D. Guerreiro, Renaud Vincentelli, and Carlos M.G.A. Fontes

Abstract

Gene synthesis is becoming an important tool in many fields of recombinant DNA technology, including recombinant protein production. De novo gene synthesis is quickly replacing the classical cloning and mutagenesis procedures and allows generating nucleic acids for which no template is available. Here, we describe a high-throughput platform to design and produce multiple synthetic genes (<500 bp) for recombinant expression in *Escherichia coli*. This pipeline includes an innovative codon optimization algorithm that designs DNA sequences to maximize heterologous protein production in different hosts. The platform is based on a simple gene synthesis method that uses a PCR-based protocol to assemble synthetic DNA from pools of overlapping oligonucleotides. This technology incorporates an accurate, automated and cost-effective ligase-independent cloning step to directly integrate the synthetic genes into an effective *E. coli* expression vector. High-throughput production of synthetic genes is of increasing relevance to allow exploring the biological function of the extensive genomic and meta-genomic information currently available from various sources.

Key words Gene design, Gene synthesis, High-throughput (HTP), Codon optimization, PCR assembly, Ligase-independent cloning (LIC)

1 Introduction

Synthetic biology is quickly becoming one of the most attractive areas of research, thanks to recent developments in gene synthesis technologies. In combination with intelligent gene design, gene synthesis is emerging as a valuable tool to boost recombinant protein expression. Intelligent gene design allows optimizing codon usage to the recombinant host system, thus promoting the effective operation of the heterologous cellular translational machinery. In addition, gene synthesis creates de novo DNA molecules, which enables exploring the large genomic and meta-genomic sequence data information for which tangible DNA is not accessible. Today, the emergent field of synthetic biology is generating insatiable

Lucília Domingues (ed.), *PCR: Methods and Protocols*, Methods in Molecular Biology, vol. 1620,
DOI 10.1007/978-1-4939-7060-5_7, © Springer Science+Business Media LLC 2017

demands for synthetic genes, which far exceed existing gene synthesis capabilities. Thus, it is crucial to develop simple gene synthesis technologies that will allow synthesizing multiple artificial DNA constructs of any size or sequence using rapid, accurate, cost-effective and high-throughput (HTP) protocols.

Methods for de novo chemical synthesis of DNA have been refined over the last years. A variety of these methodologies are based on the assembling of oligonucleotides into complete genes using PCR assembly [1, 2] or template-directed ligation [3]. Recently, improved methods have been described, which incorporate significant simplifications over earliest strategies [4–7]. In general, the entire gene synthesis process involves seven major steps: (1) gene design and sequence optimization; (2) oligonucleotides design and synthesis; (3) gene assembly using PCR-based strategies; (4) insertion of the generated nucleic acid into a cloning vector; (5) plasmid purification; (6) sequence verification and error removal; and (7) synthetic gene product preparation for downstream applications. One important drawback of current protocols is the low-throughput and high-cost associated with present technologies [8]. The establishment of HTP gene synthesis approaches should allow obtaining 96, or more, synthetic genes simultaneously and requires high-fidelity oligonucleotides and DNA polymerases [7] used in optimized protocols. HTP gene synthesis pipelines are powerful tools to increase the throughput of artificial gene synthesis while decreasing production costs, thus contributing to create, in a simple manner, new biological products that can potentially transform biomedical research.

This chapter reports an innovative HTP strategy for gene synthesis that was used to produce thousands of genes encoding venom peptides for efficient expression in *Escherichia coli*. This platform was designed to produce small synthetic genes (up to 500 bp) following simple and accurate HTP protocols (multiples of 96 genes). The pipeline is totally automated (using a Tecan liquid handling robot) but may be implemented manually. The developed strategy avoids additional steps for the removal of errors from synthetic genes, such as site-directed mutagenesis or enzyme treatments using proteins involved in the recognition of mismatches within DNA sequences [9–11]. Briefly, the HTP gene synthesis protocol described here commences by designing the genes (*see* Fig. 1) by back-translating the peptide sequence and optimizing

Fig. 1 (continued) 24-DW plates containing LB medium supplemented with kanamycin (A′–D′). On day 4, synthetic genes cloned into pHTP4 vector are isolated by DNA plasmid purification and sequenced. Finally, DNA sequences are checked for the presence of errors using NZYMulti Alignment tool, which allows analysing all data simultaneously. ◆ highlights bioinformatics tools used for HTP gene and oligonucleotide design and sequencing analysis (NZYTech, Ltd.) ♣ indicates steps that can be performed in an automated format using a liquid handling robot containing a vacuum system. # represents checkpoints for control the efficiency of: #1—PCR assembly; #2—PCR clean-up; #3—Plasmid purification

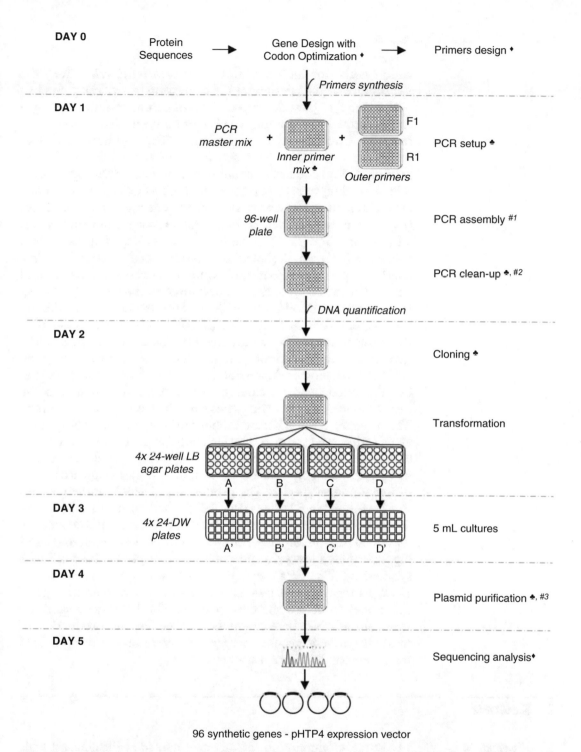

Fig. 1 Schematic representation of the HTP gene synthesis platform used for the production of multiples of 96 synthetic genes. This pipeline starts with gene design and codon optimization: multiple peptide sequences are back-translated and optimized for expression in *E. coli*, using the ATGenium codon optimization algorithm. Subsequently, oligonucleotides are designed using the NZYOligo designer. Day 1 starts with inner primer mix preparation, followed by PCR assembly and PCR cleanup. On day 2, synthetic genes are directly cloned into the *E. coli* pHTP4 expression vector using the NZYEasy Cloning & Expression kit (NZYTech, Ltd.). Cloning products are transformed into *E. coli* DH5α cells and distributed into four 24-well LB agar plates supplemented with kanamycin (A–D). On day 3, an isolated colony from each well is inoculated into the correspondent well of four

codon usage for high levels of expression in *E. coli*. Given the importance of gene design for the expression of artificial genes in a host system [12], we developed a robust codon optimization algorithm, termed ATGenium, which uses an optimized codon usage table for *E. coli* or any other organism. This algorithm allows to select the best DNA sequence from an enormous number of possible DNA sequences encoding the desired heterologous protein, according to several parameters such as GC content or presence of repetitive sequences, absence of gene regulatory sequences (e.g., promoters, activators, or operators), or maximizing codon adaptation index (CAI). In addition, we developed an algorithm, termed NZYOligo designer, to simultaneously design multiple overlapping oligonucleotides required to produce the desired genes. The next step of this method involves the assembly of overlapping oligonucleotides by PCR, where the oligonucleotides are used as templates by a high-fidelity DNA polymerase to assemble the full-length gene in a single step (*see* Fig. 2). Next, the artificial gene is directly cloned into the final destination vector using the NZYEasy Cloning & Expression System (NZYTech, Ltd.), a versatile ligase-independent cloning (LIC) method that allows the direct cloning of synthetic genes into *E. coli* expression vectors. This represents a significant simplification of the technology as it avoids additional steps to transfer synthetic genes from cloning to expression vectors. This platform was recently used to produce 4992 synthetic genes encoding venom peptides originated from different animal species. The production of 4992 synthetic genes was performed in 52 plates (96-well format) with a throughput of 4 × 96-well plates per week. The HTP gene synthesis pipeline described here displays high efficiencies of PCR assembly and cloning, revealing a low error rate of 1.06 error per kilobase of DNA synthesized. The data also revealed the strong robustness of the HTP platform since it was only necessary to screen an average of 1.3 clones to obtain the correct gene. The HTP gene synthesis strategy described here might be applied for the production of multiple synthetic genes either from different origins or optimized for expression in different host systems.

2   Materials

2.1 Polymerase Chain Reaction (PCR)

1. Primers: oligonucleotides are synthesized at the smallest scale (5 µM), with desalting purification, dissolved in 10 mM Tris–HCl buffer, pH 8.0, and stored at −20 °C.

2. 96-well ELISA plates.

3. Enzymes: KOD Hot Start DNA polymerase (1 U/µL, EMD-Millipore).

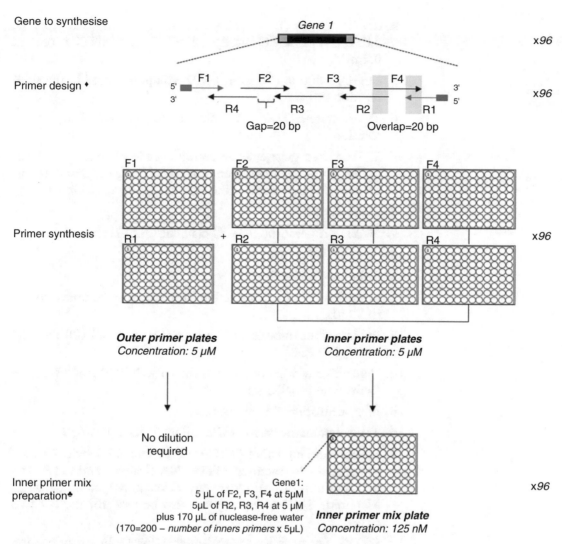

Fig. 2 Schematic diagram representing primer design, arrangement of plates for primer synthesis, and dilution to be used for HTP gene synthesis production. After gene design, the 96 genes to synthesize are organized in a 96-well format according to gene length. For each gene, a well position is defined and it is kept until the last step of the HTP gene synthesis pipeline. This diagram presents the different steps involved in primer design, primer synthesis and inner primer mix preparation for Gene 1 (340 bp), which is located in well position A01. Each oligonucleotide is represented as an *arrow*, *black arrows* correspond to internal (inner) oligonucleotides, while external (outer) oligonucleotides are denoted as *grey arrows*. Each strand of the desired gene is dissected into 60 bp overlapping oligonucleotides spaced by 20 bp (named gap), with 20 bp overlap regions between forward and reverse oligonucleotides (highlighted in *light grey*). Two outer oligonucleotides incorporate a 16 bp vector-complementary region at the 5′-end (highlighted in *dark grey rectangles*) to ensure an efficient LIC reaction. Following primer design, all oligonucleotides are organized by primer name in different 96-well plates. The total number of primer plates required is equal to the number of primers needed to synthesize the desired gene (i.e., the synthesis of Gene 1 requires a total of eight primers that are distributed along eight different primer plates). The inner primer mix will pool together in same well all inner primers at an equal concentration of 125 nM. To do this and for Gene 1, a mixture of 5 μL of each inner primer (F2, F3, F4, R2, R3, and R4 primers) and 170 μL of nuclease-free water is prepared in well position A01 of a new 96-well ELISA plate, named inner primer mix plate. ◆ highlights bioinformatics tool used for HTP oligonucleotide design (NZYOligo designer). ♣ indicates steps that can be performed in an automated format using a liquid handling robot containing a vacuum system

4. dNTP solution (2 mM, EMD-Millipore): 2 mM dATP, 2 mM dTTP, 2 mM dCTP, and 2 mM dGTP. Each dNTP is used at 0.2 mM.

5. $MgSO_4$ solution (25 mM, EMD-Millipore), used in the PCR reaction at 1.5 mM.

6. Tris-Acetate-EDTA (TAE) Buffer 50×, pH 8.3. Use as a 1× solution.

7. TAE-agarose gels: agarose routine grade diluted in 1× TAE buffer, plus GreenSafe Premium (NZYTech, Ltd), cast in self-contained system for routine agarose gel electrophoresis.

8. DNA ladder.

9. NZYDNA Clean-up 96 well plate kit (NZYTech, Ltd).

10. 96-well PCR plates.

11. Adhesive PCR seals.

12. Multichannel pipettes suitable for dispensing volumes from 5 to 50 µL.

13. PCR machine suitable for 96-well plates (such as T100 Thermal cycler, Bio-Rad).

14. Centrifuge with rotor adapted for 96-well PCR plates (such as Centrifuge 5810R, Eppendorf).

15. Microcentrifuge for microtubes.

16. Transilluminator (such as ChemiDoc™ XRS, Bio-Rad).

17. Liquid handling robot (TECAN Freedom EVO series) is used to set up PCR reactions, HTP DNA clean-up and HTP plasmid purification. Alternatively, a vacuum system (such as Manifold, EMD-Millipore) can also be used for the last two steps.

18. UV-Vis Spectrophotometer suitable for DNA quantification (such as Microplate reader, Thermo Scientific Multiskan GO UV/Vis microplate spectrophotometer with µDrop Plate, Thermo Scientific) or NanoVue (GE Healthcare Life Sciences).

2.2 Cloning and Transformation

1. Cloning system: All cloning reactions are performed using the NZYEasy Cloning & Expression kit (NZYTech, Ltd). This system allows direct cloning of synthetic genes into expression vectors, avoiding additional steps of transferring genes from cloning to expression vectors. In the current strategy, genes are directly inserted into pHTP4 *E. coli* expression vector. This vector is designed for direct cloning and high-level expression of peptide sequences fused with a disulfide isomerase (DsbC) protein. The vector carries the resistance gene for kanamycin and includes an N-terminal 6-histidine tag which allows recombinant protein purification through IMAC (ion metal affinity chromatography).

2. Multichannel pipettes with variable span (suitable with volumes from 5 to 10 µL), automatic multichannel pipettes (such as Matrix Equalizer Pipettes-125 µL, Thermo Scientific) which are used to dispense reagents into a 24- and 96-well format, and 100 mL disposable reagent reservoirs.

3. 96-well PCR plates.

4. Adhesive PCR seals.

5. PCR machine suitable for 96-well PCR plates.

6. Competent cells: Chemical competent DH5α *E. coli* strain.

7. Antibiotic: kanamycin (50 mg/mL in water), store stock solution at −20 °C and use a 1/1000 dilution.

8. LB Broth: dissolve 10 g tryptone, 5 g yeast extract and 10 g NaCl in 950 mL of ultrapure water. Adjust the pH to 7.0 using NaOH and add ultrapure water up to 1 L. Mix well and dispense into appropriate containers. Sterilize in autoclave at 121 °C for 15 min.

9. LB agar: Dissolve 10 g tryptone, 5 g yeast extract and 10 g NaCl in 950 mL of ultrapure water. Adjust the pH to 7.0 using NaOH. Add 20 g of agar and adjust the volume to 1 L with ultrapure water. Mix well, dispense into appropriate containers and autoclave.

10. LB agar plates: Melt slowly a bottle of LB agar in a microwave. Once cooled, add the required antibiotic (50 µg/mL kanamycin). Distribute 1.5 mL of LB agar to each well of 24-well sterile tissue culture plates.

11. 24-well LB agar plates: 24-well sterile tissue culture plates (Greiner Bio-One). For 96 transformations, use 4 × 24-well sterile tissue culture plates.

12. SOC medium: Dissolve 20 g tryptone, 5 g yeast extract, 0.5 g NaCl in 950 mL of ultrapure water, and add 4.5 mL of 0.5 M KCl. Mix well and adjust to 1 L the medium volume. Sterilize in autoclave at 121 °C for 15 min. Once cooled, add 9 mL of 2 M $MgCl_2$ hexahydrate and 20 mL of 1 M (20%) glucose. The prepared medium should be stored at 2–8 °C.

13. Multidispenser (such as MINILAB 201 dispenser, HTL) and correspondent 50 and 25 mL syringes.

14. 24-deep-well plates (24-DW, Greiner Bio-One). For 96 cultures, use 4 × 24-DW plates.

15. Air-O-top seals (4-titude Ldt or similar).

16. Centrifuge with rotor for 24-DW plates (such as Avanti J-E centrifuge, Beckman Coulter).

17. NZYMiniprep 96 well plate kit (NZYTech, Ldt).

18. 96-well ELISA plates.

19. Liquid handling robot (TECAN Freedom EVO series).

20. Water bath set at 42 °C.

21. Shaking incubator set to 37 °C.

22. Plate incubator set to 37 °C.

2.3 Bioinformatics Tools

The following bioinformatics tools are intellectual property of NZYTech Ltd., and are available upon request.

1. ATGenium codon optimization software—High-throughput gene optimization.

2. NZYOligo designer program—High-throughput primer design.

3. NZYMulti Alignment software—High-throughput sequencing data analysis.

3 Methods

An integrated gene synthesis platform was developed for the efficient production of small synthetic genes (<500 bp) based on oligonucleotide assembly by PCR (*see* **Note 1**). This platform combines automation, simplicity and robustness, while decreasing the error rate associated with conventional gene synthesis methods. Initial experiments (not shown) allowed defining the most appropriate PCR assembly protocol, which is described below. The whole procedure is designed to be performed in 5 days and consists in seven main steps: (1) gene design with codon optimization, (2) primer design and synthesis, (3) PCR setup and assembly, (4) PCR clean-up, (5) gene cloning and transformation, (6) plasmid purification, (7) gene sequencing and sequence analysis (Fig. 1). All steps described below are defined to synthesize 96 genes simultaneously, thus using 96-well plates, although the whole process can be applied to produce smaller or higher number of synthetic genes (e.g., 24, 48, 72, 96, 192, or multiples). These protocols can be performed manually using multichannel pipettes and a vacuum system (such as the manifold, EMD-Millipore) or fully automated, as described here, using a liquid handling robot containing a vacuum system (such as the TECAN Freedom EVO series). This HTP gene synthesis platform was validated to successfully obtain 384 synthetic genes in 1 week, using an automated system.

3.1 Gene Design Using ATGenium Codon Optimization Algorithm

To maximize levels of expression of synthetic genes, DNA sequences must be optimized according to the host system codon usage. Genes were therefore designed using ATGenium codon optimization algorithm, by backtranslating corresponding peptide sequence and by optimizing codon usage for high levels of expression in *E. coli* (*see* **Note 2**). Codons were selected randomly using a Monte Carlo approach according to *E. coli* codon usage (*see* **Note 3**) of highly expressed genes, excluding naturally rare codons. Genes were designed to have an optimal GC content (between 40% and 60%) and a codon adaptation index (CAI) value higher than 0.8. Presence of G/C islands, which could promote frame-shifting, was minimized by selectively avoiding runs of consecutive G and/or C greater than six nucleotides. In addition, no contiguous strings of nucleotides longer than five nucleotides were allowed. Genes were engineered to ensure the absence of *E. coli* regulatory sequences such as promoters, activators, or operators, and unwanted restriction sites.

3.2 Oligonucleotide Design Using NZYOligo Designer Program

The primer pool required to synthesize 96 artificial genes was designed using the NZYOligo designer program, which allows working simultaneously with multiple DNA sequences. The algorithm used for oligonucleotide design allows defining primer lengths, gap regions, overlapping regions by defining start and end positions, and introducing engineered 5′- and 3′-end sequences. The DNA sequence of each gene is used as template to design the assembly oligonucleotides by dividing the entire sequence into overlapping primers with defined lengths. The external oligonucleotides, termed outer primers, correspond to the external forward and reverse primers and internal oligonucleotides (termed inner primers) are usually present in higher number than outer primers. Primers used for PCR amplification had typically a length of 60 bp, an overlap region of 20 bp between forward and reverse primers, as well as a gap of 20 bp (*see* Fig. 2 and **Note 4**). Outer primer sequences included a 16-bp overhangs on the 5′-terminus of both forward and reverse primers (*see* Table 1) to allow ligase-independent cloning into the pHTP4 *E. coli* expression vector (*see* **Notes 5** and **6**).

In large projects, the arrangement of genes in a 96-well format should be done according with gene size to facilitate the preparation of inner oligonucleotide mixtures. This enables keeping together the genes that are assembled from a similar number of oligonucleotides. Plate organization should be defined after gene design. At this stage, a well position is defined and fixed for each gene to synthesize by PCR assembly. For each gene, the same well position is kept for respective primers and for subsequent steps.

Table 1
Overhang sequences included in the outer primers that allow cloning into the pHTP4 vector

Outer primer overhang	Sequence
Forward	5′-TCAGCAAGGGCTGAGG…-3′
Reverse	5′-TCAGCGGAAGCTGAGG…-3′

Thus, the gene synthesis procedure is more efficient when primers are ordered in separate 96-well plates, according to primer name (*see* Fig. 2). Thus, the total number of primer plates required is equal to the number of primers needed to synthesize the entire gene length (*see* Fig. 2). We recommend obtaining all oligonucleotides at 5 μM in nuclease-free water, without purification (*see* **Notes** 7 and **8**).

3.3 Gene Assembly

1. For each gene to be synthesized, prepare an inner primer mix at 125 nM containing all inner oligonucleotides. For this, add x μL of nuclease-free water ($x = 200$ μL $- n \times 5$ μL, n—number of inner primers that depends on the length of the target gene) and add 5 μL of each inner primer at 5 μM to the corresponding well of the inner primer mix plate (*see* Fig. 2 and **Note 9**). Seal the plate with an adhesive PCR seal and centrifuge to spin down.

2. A single PCR reaction is performed in a final volume of 50 μL. On ice, prepare a PCR master mix sufficient for 96 PCR reactions (we recommend preparing a PCR master mix in excess, for example for 104 reactions) by combining the reagents specified in Table 2 (*see* **Note 10**). Mix well the solution and centrifuge to spin down.

3. Using a multichannel pipette, distribute 26 μL of PCR master mix into each well of a 96-well PCR plate.

4. Using a multichannel pipette, add 8 μL of inner primer mix at 125 nM, followed by 8 μL of each outer primers at 5 μM (F1 and R1 primers) into the corresponding well of the 96-well PCR plate containing the PCR master mix. Seal the plate using an adhesive PCR seal and centrifuge to spin down. The two outer primers are used at a final concentration of 800 nM while the inner primers are pooled together in an equimolar mixture to achieve the final concentration of 20 nM in PCR reaction.

Table 2
PCR components used to perform 96 PCR assembly reactions

Components	Volumes for 1 rx. (μL)	Volumes for 104 rxs. (μL)
10× Reaction buffer	5	520
25 mM $MgSO_4$	3	312
2 mM dNTP mixture	5	520
KOD Hot Start DNA polymerase	1	104
H_2O	12	1248

5. Place the 96-well PCR plate into a thermocycler and perform PCR assembly using the following cycling parameters:

95 °C for 2 min: 1 cycle

95 °C for 20 s; 60 °C for 8 s; 70 °C for 3 s: 26 cycles

4 °C: hold

6. Checkpoint #1: confirm the efficiency of PCR assembly by analysing a sample of eight out of the 96 reactions (one column of the 96-well PCR plate, for example), through agarose gel (1.5% w/v) electrophoresis. Prepare a mixture of 5 μL of each PCR product plus 1 μL of 6× NZYDNA loading dye and run 5 μL of the mixture against a DNA ladder on agarose gel. Agarose gel electrophoresis should reveal eight clear bands with the correct gene size (*see* **Note 11**).

3.4   PCR Clean-Up

1. Proceed with silica-column purification of the PCR product using NZYDNA Clean-up 96 well plate kit. This procedure will remove dNTPs, unused primers and other impurities. Purify the PCR products following the manufacturer's instructions using the vacuum automated format (*see* **Note 12**).

2. Following PCR clean-up, quantify PCR products using a UV/Vis microplate spectrophotometer, by measuring absorbance at 260 nm. Determine an average concentration for 16 PCR products (in ng/μL) and use this value to calculate the amount of PCR product to use per cloning reaction, as explained below.

3. Checkpoint #2: confirm the efficiency of the PCR clean-up step by analysing a sample of eight out of the 96 reactions (one column of the 96-well PCR plate, for example), through agarose gel (1.5% w/v) electrophoresis (prepare and run samples as explained above). Agarose gel electrophoresis should reveal eight clear bands with the correct gene size (*see* **Note 11**).

3.5   Cloning and Transformation

Cloning of purified PCR products into the pHTP4 expression vector should be performed using the NZYEasy Cloning & Expression System that follows a LIC technology (*see* **Notes 13** and **14**). To achieve high cloning efficiencies, we recommend using the following equation which allows calculating the amount of PCR product to be used per cloning reaction.

$$PCR \text{ product required} \left(ng\right)^{*} = PCR \text{ product length} \left(bp\right)^{*} \times 0.083$$

*average concentration and PCR product length previously calculated for 16 genes

1. A single cloning reaction is performed in a final volume of 10 μL. On ice, prepare a cloning master mix sufficient for 96 reactions (we recommend to prepare a cloning master mix in

Table 3
Cloning reaction components used to perform 96 cloning reactions

Components	Volumes for 1 rx. (μL)	Volumes for 104 rxs. (μL)
10× Reaction buffer	1	104
pHTP4 vector	1	104
NZYEasy enzyme mix	0.5	52
Purified PCR product	x	$x \times 104$
H₂O	up to 10 μL	up to 1040 μL

x: average volume of purified PCR product to use in one cloning reaction. Do not include in preparation of cloning master mix

excess, for example for 104 reactions) by combining the reagents specified in Table 3, excluding the purified PCR product that will be added following master mix distribution into the 96-well PCR plate.

2. Mix the master mix and spin down. Calculate the volume of master mix to aliquot per well as follows:

$$\text{Volume of master } mix\,(\mu L)$$
$$= 10 - \text{average volume of } PCR \text{ product}\,(\mu L)$$

3. Using a multichannel pipette, distribute the cloning master mix into a 96-well PCR plate.

4. Using a multichannel pipette, add x μL of each purified PCR product into the 96-well PCR plate. Seal the plate with an adhesive PCR seal and centrifuge briefly to collect the reaction components.

5. Place the 96-well PCR plate into the thermocycler and perform the cloning reaction using the following parameters:

   37 °C for 60 min.

   80 °C for 10 min.

   30 °C for 10 min.

   4 °C, hold.

6. Centrifuge briefly to collect the reaction components and store samples on ice or at −20 °C for subsequent transformation.

7. With a multidispenser, prepare 4 × 24-well LB agar plates containing LB agar supplemented with 50 μg/mL kanamycin (*see* **Note 15**).

8. Place a new 96-well PCR plate on ice. Thaw 8 × 400 μL of DH5α *E. coli* chemically competent cells on ice and then aliquot 25 μL of competent cells into each well of the

96-well PCR plate, using an automatic multichannel pipette (*see* **Note 16**).

9. Using a multichannel pipette, add 5 μL of cloning reaction to the competent cells and ensure that the recombinant plasmids are dispensed into the cells, but do not mix by pipetting. Cover the plate with an adhesive PCR seal. Incubate the plate on ice for 30 min (*see* **Note 17**).

10. Heat-shock the cells at 42 °C for 40 s, then transfer the 96-well PCR plate back to ice for 2 min (*see* **Note 17**).

11. Add 100 μL of SOC medium into each well, cover with a new adhesive PCR seal to prevent contamination. Incubate at 37 °C for 90 min in a shaking incubator with vigorous agitation.

12. Using an automatic multichannel pipette, dispense 60 μL of transformed cells onto the previously prepared 4 × 24-well LB agar plates (*see* **Note 18**). In order to spread the cells, place the four plates in a shaking incubator set to 37 °C, for 60 min, with gentle shacking. Invert the plates and transfer them to the plate incubator, where they should be left overnight at 37 °C.

13. After overnight incubation, inspect LB agar plates for the presence of colonies. Record the number of colonies observed for each of the 96 transformation reactions.

14. Prepare 4 × 24-deep-well plates containing 5 mL of LB liquid media supplemented with 50 μg/mL of kanamycin. Inoculate one single colony per transformation and seal the plates with air-O-top seals. Be aware to avoid cross-contaminations. Incubate cultures at 37 °C for ~16 h, with gentle agitation.

3.6 Plasmid Purification

1. Using a centrifuge with a rotor for 24-DW plates, harvest cells at 1500 ×*g* for 15 min at 4 °C. Discard the supernatant into a waste container with sodium hypochlorite for decontamination before disposal. Tap the plates upside down onto absorbent paper to remove any remaining liquid medium.

2. Purify plasmid DNA from the bacterial pellets using the NZYMiniprep 96 well plate kit following the manufacturer's instructions for vacuum automated format (*see* **Note 19**).

3. Checkpoint #3: quantify DNA concentration using a UV/Vis microplate spectrophotometer, by measuring absorbance at 260 nm. Additionally, analyse a sample of eight out of the 96 DNA preparations (one column of the 96-well ELISA plate, for example), through agarose gel (0.8% w/v) electrophoresis, to confirm the concentration of plasmid DNAs. Agarose gel electrophoresis should reveal a clear band with the correct plasmid size (*see* **Note 11**).

4. Store plasmid DNA at −20 °C.

3.7 Quality Control

1. Confirm the integrity of the gene sequences by Sanger Sequencing. The artificial genes should be sequenced in both directions. Quality control should ensure consistency of each synthetic gene with the defined sequence (*see* **Note 20**).

2. Sequencing results are analysed using the NZYMulti Alignment software or by performing a sequence alignment between template DNA sequences and sequencing results of the 96 plasmids (*see* **Notes 21** and **22**).

4 Notes

1. The method described here is optimized for the simultaneous synthesis of genes with lengths up to 500 bp. This protocol can be adapted to synthesize DNA fragments longer than 500 bp. In this case, we recommend to first separate the target DNA into 500 bp fragments with 20–30 bp overlaps between adjacent DNA fragments. Each DNA fragment should be synthesized using this protocol and the entire gene can be assembled in a second PCR reaction. This strategy will decrease the error rate associated with the classical gene synthesis protocols used for large genes that require a large number of oligonucleotides.

2. ATGenium codon optimization algorithm provides an optimized *E. coli* codon usage table in order to enhance the expression levels of recombinant proteins.

3. If a different host system is required for protein expression, ATGenium codon optimization algorithm is compatible with codon usage tables from different hosts.

4. We recommend designing oligonucleotides with a length between 50 and 60 bp, which is, for most commercial DNA oligonucleotides providers, the maximum synthetic length that ensures high fidelity. This length is probably the best choice combining optimal performance and reduced cost.

5. In case DsbC fusion tag (included in the pHTP4 expression vector) removal is required following protein purification, the sequence encoding a protease recognition site (such as Tobacco Etch Virus, TEV) should be included immediately after the forward overhang and in frame with the artificial gene sequence.

6. pHTP4 expression vector contains two histidine tags (N- and C-terminal). If a C-terminal His-tag is not desired, include a stop codon in frame with the gene sequence.

7. Oligonucleotides can be synthesized at any concentration as long as the final inner and outer primers concentrations are kept at 20 and 800 nM, respectively, in the PCR reaction.

8. The high cost associated with oligonucleotide purifications (such as HPLC, high performance liquid chromatography, or

PAGE, polyacrylamide gel electrophoresis) suggest that these additional steps should be avoided during oligonucleotide synthesis. From our experience, additional purifications did not significantly improve the success of this gene synthesis protocol. Oligonucleotides can be purchased from major manufacturers. Evaluate their quality by synthesizing genes using oligonucleotides from different sources and checking the integrity of the resulting artificial nucleic acids.

9. In order to avoid pipetting errors, a robotic system is recommended to prepare the inner primer mix.

10. A high-fidelity DNA polymerase is strongly recommended to minimize errors introduced during gene assembly by PCR. We recommend KOD Hot Start DNA polymerase (EMD-Millipore).

11. Checkpoints: if no bands are visible on agarose gel, we recommend analysing more samples in order to verify the success of the PCR assembly, PCR clean-up or plasmid purification. In case of negative results, consider that the respective step was not correctly performed.

12. To recover DNA, we recommend adding 80 µL of Elution Buffer or nuclease-free water (pH 8.5) to each well and incubate for 1–3 min at room temperature in order to increase elution efficiency.

13. One of the main advantages of this system is the possibility of cloning synthesized genes directly into *E. coli* expression vectors. Unlike other commercial kits, no additional sub-cloning steps are required, which saves time and decreases costs.

14. In case of the best expression vector is unknown, we recommend the following strategy: clone the genes of interest directly into the pHTP0 cloning vector (ampicillin resistant). Subsequently, transfer genes to pHTP expression vectors (NZYTech, Ltd.) which contain different solubility tags and kanamycin resistance, thus generating different expression clones of the same gene. Check the levels of recombinant protein expression and solubility using conventional protocols. This strategy allows selecting the most appropriate solubility tag for each protein.

15. LB agar plates can be prepared in advance and stored for up to 2 weeks at 4 °C. Before usage, they should be prewarmed and dried at 37 °C, by leaving them in a plate incubator placed upside down.

16. It is important to use high efficiency DH5α competent cells for the transformation. If you fail to get colonies, repeat the transformation using 50 µL of cells plus the remaining 5 µL of cloning reaction.

17. Transformation protocol can be performed on a water bath or on a PCR machine.

18. When using an automatic multichannel pipette to dispense transformed cells onto 24-well LB agar plates, care should be taken to prevent cross-contamination.

19. To recover plasmid DNA, we recommend adding 120 μL of Elution Buffer or nuclease-free water (pH 8.5) to each well, and incubate for 1–3 min at room temperature in order to increase elution efficiency.

20. For sequencing pHTP expression vectors, we recommend using T7 universal primer (5′-TAATACGACTCACTATAGG G-3′) and T7 terminator primer (5′-GCTAGTTATTGCTC AGCGG-3′).

21. The error rate of the large-scale method described here is 1.06 mutations per kilobase of DNA synthesized (single base deletions are more frequent, followed by substitutions and insertions). In case the DNA sequence is not 100% identical to the designed sequence, screen a second or a third colony (an average of 1.3 clones are necessary to obtain the correct gene).

22. The low error rate of this gene synthesis method avoids additional steps for the removal of errors. Since the identification of 100% correct genes is performed by screening a maximum of three colonies, the labor required for the selection and validation of recombinant clones is reduced.

References

1. Hoover DM, Lubkowski J (2002) DNAWorks: an automated method for designing oligonucleotides for PCR-based gene synthesis. Nucleic Acids Res 30:e43

2. Stemmer WP, Crameri A, Ha KD et al (1995) Single-step assembly of a gene and entire plasmid from large numbers of oligodeoxyribonucleotides. Gene 164:49–53

3. Strizhov N, Keller M, Mathur J et al (1996) A synthetic cryIC gene, encoding a bacillus thuringiensis delta-endotoxin, confers Spodoptera resistance in alfalfa and tobacco. Proc Natl Acad Sci U S A 93:15012–15017

4. Xiong A-S, Yao Q-H, Peng R-H et al (2004) A simple, rapid, high-fidelity and cost-effective PCR-based two-step DNA synthesis method for long gene sequences. Nucleic Acids Res 32:e98–e98. doi:10.1093/nar/gnh094

5. Xiong A-S, Yao Q-H, Peng R-H et al (2006) PCR-based accurate synthesis of long DNA sequences. Nat Protoc 1:791–797. doi:10.1038/nprot.2006.103

6. Gordeeva TL, Borschevskaya LN, Sineoky SP (2010) Improved PCR-based gene synthesis method and its application to the Citrobacter freundii phytase gene codon modification.

J Microbiol Methods 81:147–152. doi:10.1016/j.mimet.2010.02.013

7. Wu G, Wolf JB, Ibrahim AF et al (2006) Simplified gene synthesis: a one-step approach to PCR-based gene construction. J Biotechnol 124:496–503. doi:10.1016/j.jbiotec.2006.01.015

8. Tian J, Ma K, Saaem I (2009) Advancing high-throughput gene synthesis technology. Mol BioSyst 5:714. doi:10.1039/b822268c

9. Ma S, Saaem I, Tian J (2012) Error correction in gene synthesis technology. Trends Biotechnol 30:147–154. doi:10.1016/j.tibtech.2011.10.002

10. Saaem I, Ma S, Quan J, Tian J (2012) Error correction of microchip synthesized genes using surveyor nuclease. Nucleic Acids Res 40:1–8. doi:10.1093/nar/gkr887

11. Sequeira AF, Guerreiro CIPD, Vincentelli R, Fontes CMGA (2016) T7 endonuclease I mediates error correction in artificial Gene synthesis. Mol Biotechnol. doi:10.1007/s12033-016-9957-7

12. Welch M, Govindarajan S, Ness JE et al (2009) Design parameters to control synthetic Gene expression in *Escherichia coli*. PLoS One 4:e7002. doi:10.1371/journal.pone.0007002

Chapter 8

Colony PCR

Flávio Azevedo, Humberto Pereira, and Björn Johansson

Abstract

Escherichia coli and *Saccharomyces cerevisiae* are currently the two most important organisms in synthetic biology. *E.coli* is almost always used for fundamental DNA manipulation while yeast is the simplest host system for studying eukaryotic gene expression and performing large scale DNA assembly. Yeast expression studies may also require altering of the chromosomal DNA by homologous recombination. All these studies require the verification of the expected DNA sequence and the fastest method of screening is colony PCR, which is direct PCR of DNA in cells without prior DNA purification. Colony PCR is hampered by the difficulty of releasing DNA into the PCR mix and the presence of PCR inhibitors. We hereby present one protocol for *E. coli* and two protocols for *S. cerevisiae* differing in efficiency and complexity as well as an overview of past and possible future developments of efficient *S. cerevisiae* colony PCR protocols.

Key words PCR, Colony, Yeast, *Saccharomyces cerevisiae*, *Escherichia coli*, Direct lysis

1 Introduction

Colony or whole cell PCR is the direct PCR amplification of target sequences from cells without prior isolation or purification of DNA. Colony PCR is possible if enough cells lyse as a consequence of the high temperature in the initial template denaturation step alone or in combination with extra procedures to make DNA more accessible. The material containing the cells or the cells themselves must also not present PCR inhibition to an extent that prevents PCR amplification. The advantage of colony PCR over using purified DNA is mainly speed and cost, as the time consuming DNA extraction step is omitted; but omitting DNA purification and thereby minimizing sample handling can also increase sensitivity if the starting material is limiting, as it might be in, for example, forensic applications. Very low amounts of starting material may prohibit DNA purification as all purification procedures are associated with a loss. Less sample handling also lowers the risk of cross contamination of samples, an important point since PCR is a sensitive technique, which may be prone to false positive detection.

Lucília Domingues (ed.), *PCR: Methods and Protocols*, Methods in Molecular Biology, vol. 1620,
DOI 10.1007/978-1-4939-7060-5_8, © Springer Science+Business Media LLC 2017

The need for detecting the presence or absence of specific sequences within cells is routinely needed in a wide range of disciplines, such as clinical microbiology, genetic engineering and forensic sciences. The most common application of colony PCR in genetic engineering is probably the amplification of ligation product sequences within *Escherichia coli* transformants after cut and paste cloning. This procedure is straightforward with few associated problems. The first report on *E. coli* colony PCR describe the resuspension of one colony in half a mL of water and subsequent boiling for 5 min [1]. After centrifugation for 2 min at maximum speed in a microcentrifuge ($15 - 20,000 \times g$), 5 µL of the supernatant was used for PCR. The authors later succeeded in using *E. coli* directly as template without prior dilution or boiling. Few variations of this simple protocol have been published, indicating that it is generally applicable with a reasonable rate of success. Current iteration of the protocol simply involves adding a small amount of an *E. coli* colony to a PCR reaction which is thereafter handled as if an amplification from pure DNA.

Another common application of colony PCR is the analysis of *Saccharomyces cerevisiae* transformants after genetic engineering or DNA assembly experiments. *S. cerevisiae* has the capacity to assemble large and complex constructs through homologous recombination in one step [2], a feature that has found many applications in the emerging field of synthetic biology [3].

Colony PCR from *S. cerevisiae* is a non trivial technique, which is evident from the myriad of available techniques, both published with peer review and available online (*see* [4] for a compilation). This indicates that there may not be one protocol that is optimal for all use cases. False negative results are a general problem affecting yeast colony PCR. Factors that seem to affect yeast colony PCR efficiency are chronological age of the culture, growth phase, growth rate, size of the desired PCR product, copy number of the target sequence and media components. Fresh cultures of rapidly growing yeast, where the target amplicon is short and present in multiple copies, seem to present the least problems. Early published yeast colony PCR protocols were essentially *E. coli* protocols adapted for yeast, where yeast cells are simply added to the PCR mixture, and the cells are presumably lysed in the initial denaturation step [5].

The addition of a preincubation with a yeast lytic enzyme such as Zymolyase can improve efficiency [6]. The factor targeted by this protocol is the strong yeast cell wall, which is weakened or removed by the enzymatic treatment. The downside is enzyme cost and the addition of phosphate in the incubation buffer, which may lead to PCR inhibition by interaction with the magnesium ions in the PCR buffer. A brief treatment of cells with sodium hydroxide [7] is a method that has several potential targets. The authors suggest that the modes of action could increase cell wall permeability, dissociation of DNA from bound proteins or degradation of RNA. Additionally, sodium hydroxide might neutralize intercalated PCR inhibitors by denaturing DNA [8]. The addition of the strong anionic detergent sodium dodecyl sulfate (SDS) alone [9] or in combination with ethanol [10] or lithium acetate (LiAc) [11] have also been described as methods for achieving PCR amplification from whole yeast cells. SDS efficiently dissolves membrane lipids, but is also a potent PCR inhibitor [12]. The presence of SDS also potentially eliminates DNA-protein interactions, as SDS is used to prevent gel-shifts in electrophoresis of DNA. Ethanol would precipitate DNA as soon as it is liberated from the cells and may be a way to selectively wash away inhibitors and concentrate DNA [10]. LiAc is commonly used in yeast transformation [13], where the mode of action may be to turn the cell wall more porous [14], which probably improves cell lysis.

Physical methods as heating, boiling, grinding with glass beads or rapid freeze–thaw cycles [15] have also been employed, but may be more laborious if the number of samples is large. Glass beads in combination with the metal chelating resin Chelex 100 has been reported to permit PCR from whole yeast cells [16]. The role of the chelator is to remove metal ions necessary for nucleases, thereby protecting DNA. The use of chelating resins has also been reported to allow PCR amplification from forensic samples [17].

We have adapted one protocol for *E. coli* colony PCR (Subheading 3.1) and two different protocols for *S. cerevisiae* (Subheadings 3.2 and 3.3) that are routinely used in our laboratory. The protocol in Subheading 3.2 is very simple and rapid, involving only a short preincubation step in a microwave oven, while the protocol in Subheading 3.3 is a version of the LiAc SDS protocol [11], which is more laborious, but also in our hands more sensitive and robust than the first protocol.

2 Materials

1. Water. PCR components and other solutions should be prepared using the best available water. We routinely use double deionized water with a specific conductance of 18.2 MΩ/cm at 25 °C.

2. 2× PCR mastermix with dimethyl sulfoxide (DMSO) (Table 1). We have found it practical to prepare a two times concentrated PCR mastermix containing all components except PCR primers and template DNA, as this minimizes pipetting errors and improves consistency across PCR experiments. The PCR mastermix can be stored at −20 °C without noticeable loss of efficiency. We routinely include 1% DMSO in the final PCR mixture.

Table 1

Recipe for 1 mL twice concentrated PCR mastermix containing 2% DMSO suitable for colony PCR

Component	Volume (μL)
Water	650
Taq buffer with NH_4SO_4 (10×)	200
$MgCl_2$ (50 mM)	80
dNTPs (10 mM each)	40
DMSO (100%)	20
Taq DNA polymerase (5 U/μL)	10

Table 2

Recipe for 5× PCR compatible loading buffer

Component	Volume
25% ficoll	10 mL
Tartrazine food coloring	1 mL
Xylene Cyanol 125 mg/mL	10 μL

3. 5× PCR compatible loading buffer (Table 2). A PCR compatible loading buffer can be added directly into the PCR mix, saving post PCR pipetting steps that might potentially contaminate the laboratory. We have adopted such a loading buffer made in house to lower PCR costs. The Tartrazine food coloring is a commercial food coloring sold in grocery stores.

4. 1 M Lithium Acetate Stock Solution. This solution is a part of the LiOAc-SDS solution for protocol in Subheading 3.3. The lithium acetate solution is prepared as a 1 M stock in water. Add 10.2 g of lithium acetate dihydrate (LiOAc*2H20, Mw 102.02 g/mol) in 80 mL water and dissolve. Add water to 100 mL and autoclave.

5. SDS Stock Solution 20% (w/v). Add 10 g of SDS to 40 mL H_2O. Heat to 60 °C to dissolve the SDS. Adjust pH to 7–8 using sodium hydroxide. Adjust the volume to 50 mL with water. Do not autoclave this solution as SDS will precipitate. This solution is a component of the LiOAc-SDS solution for protocol in Subheading 3.3.

6. LiOAc-SDS Solution for protocol in Subheading 3.3. Mix 75 μL water, 20 μL of 1 M LiOAc and 5 μL of 20% (w/v) SDS

for each DNA extraction. Aliquot 100 μL in 1.5 mL microcentrifuge tubes.

7. TE buffer. Add 10 mL 1 M Tris–HCl pH 8.0 and 2 mL of 0.5 M EDTA pH 8.0 to 988 mL water. The resulting solution will be 10 mM Tris–HCl and 1 mM EDTA. Autoclave to sterilize. This buffer is used to resuspend DNA in the last step in Subheading 3.3 protocol. Other recipes of TE buffer can probably also be used.

3   Methods

Carry out all procedures at room temperature unless otherwise specified. PCR mastermixes should be kept on ice at all times, but PCR tubes can be handled at room temperature during the preparation of the mix.

3.1   E. coli Colony PCR

This protocol can be used to amplify new constructs in *E. coli* transformants. We have found it efficient to use a three primer strategy, using two vector-specific primers flanking the insertion location of the insert and one gene-specific primer, which is usually one of the primers used to amplify the insert (*see* **Note 1**). The two vector-specific primers should differ in distance to the insertion site by 200–400 bp. Using this strategy, an empty clone will produce a short PCR product corresponding to the distance between the vector-specific primers while one of two longer bands will arise depending on the orientation of the cloned insert.

1. Prepare a 1× PCR master mix containing all PCR components except template DNA. We use a homemade 2× PCR master mix containing DNA polymerase, buffer, Mg^{2+}, dNTPs, and DMSO (*see* Subheading 2) to which are added PCR primers to a final concentration of 1 μM and water. We prepare 110% of the theoretical required volume, which is calculated as the total volume of each PCR reaction times the number of clones, including two negative (no cells and empty cells) and a positive control if available. The cells are assumed to take up no volume in the calculation.

2. Prepare the appropriate number of tubes containing 1× PCR mastermix. We keep the tubes open before adding the *E. coli* cells as we have found that the proximity to a bunsen burner provide a sufficiently clean environment to avoid contamination.

3. Add a part of the *E. coli* colony to the inside of the tube, by swirling the toothpick against the wall of the tube (*see* **Notes 2–4**).

4. Transfer the remaining cells on the toothpick to fresh medium for preserving the clone and possibly preparing plasmid DNA.

5. Vortex and place the PCR tubes in a preheated thermal cycler as soon as possible.

6. Run the PCR program; time periods and temperature depend on the polymerase used, size of the expected PCR product, and melting temperature of the primers. We have found that 5-min initial denaturation (94–98 °C), 35 cycles of the main program and 5 min of post extension at 72 °C is sufficient.

7. Analyze 5–10 μL of the PCR amplification by gel electrophoresis. We add dyes to the loading buffer (*see* Subheading 2), in which case we can omit adding loading buffer to the PCR products.

3.2 S. cerevisiae Colony PCR Using a Microwave Oven

This protocol usually represents the best compromise between cost, work and success rate, and should probably be the first protocol tested for a laboratory wishing to implement *S. cerevisiae* colony PCR. We have found it to be efficient for PCR products up to 2 kb, with occasional success for products up to 3 kb in size.

1. Prepare 1× PCR master mix according to the same principles as for the *E. coli* protocol (Subheading 3.1).

2. Pick a small, well-isolated colony with a sterile toothpick or a sterile 200 μL pipette tip (*see* **Notes 3** and **5**).

3. Transfer part of the colony to the side of a PCR tube. The most common mistake is to transfer too much cell material to the tube. We usually swirl the toothpick on the inside of the tube.

4. Transfer the remaining cells on the toothpick to fresh medium.

5. Incubate the tubes for 1–2 min at full power (800–1000 W) using a stock microwave oven.

6. Cool the tubes by placing them on ice or by a 3–5-min incubation at −20 °C in a freezer.

7. Add the PCR mastermix. We use a scale of 20 μL to save on reagents. A larger scale such as 50 μL will be less sensitive to excess biomass in the PCR reaction, which might be useful for optimization.

8. Run the PCR program (*see* **Note 6**) and analyze 5–10 μL of the PCR product by gel electrophoresis.

3.3 PCR Using S. cerevisiae LiAc Permeabilized Cells

This protocol may not qualify as colony PCR, as DNA is effectively purified from the cells. However, this protocol is considerably less laborious than methods relying on any combination of glass beads, phenol, and chloroform. In our hands, this protocol has succeeded where the microwave oven protocol (Subheading 3.2) failed. This protocol has given us more stable results, especially in the hands of less experienced workers. This protocol was first described by Lõoke et al. [11].

1. Prepare one tube of 100 μL LiOAc-SDS mix for each colony.

2. Transfer a small colony from a plate using a sterile toothpick (*see* **Note 5**). The toothpick can also be used to inoculate liquid or solid medium to preserve the clone.

3. Vortex the tubes briefly and incubate at >70 °C for 10 min (*see* **Note 7**).

4. Add 300 μL of 96% ethanol and vortex briefly to precipitate DNA.

5. Spin tubes at least at 12000 rpm ($\approx$15,000 $\times$ *g*) in a microcentrifuge for 3–5 min to precipitate DNA. The cells and cell debris will co precipitate with the DNA at this point.

6. Remove the liquid by inverting the tubes.

7. Add 500 μL 70% ethanol to each tube.

8. Spin tubes like in **step 5**.

9. Remove the liquid by inverting the tubes. Try to remove as much of the liquid as possible in this step (*see* **Note 8**).

10. Resuspend the DNA in 100 μL TE buffer. Unlysed cells and cell debris will also be resuspended in this step.

11. Spin down the cells debris for 1 min at top speed in a microcentrifuge.

12. Use 1 μL of the supernatant for PCR.

13. Transfer about half of the supernatant to a fresh tube and store the DNA at −20 °C.

4   Future Development

There is substantial development of techniques for direct amplification of DNA in complex matrixes with as few or no manipulation steps involved. Rapid genetic typing of human blood or tissue and detection of human pathogens, as well as forensic science, are likely the strongest motivation for this development. It is possible that at least some of these new procedures could benefit new methods for direct colony PCR from difficult sources such as *S. cerevisiae*.

One of the most attractive recent developments are thermostable DNA polymerases engineered for higher PCR inhibitor tolerance [18]. Examples of these are the addition of DNA binding domains [19] and polymerases developed through gene shuffling or compartmentalized self-replication. The last approach has yielded DNA polymerases resistant to the potent PCR inhibitor heparin [20] and a broad range of environmentally derived inhibitors [21].

PCR enhancers are another area of development that could potentially aid colony PCR protocols. Common PCR enhancers include N,N,N-trimethylglycine (betaine), bovine serum albumin

(BSA), dithiothreitol (DTT), glycerol and DMSO. DMSO was first reported as improving Sanger DNA sequencing quality [22] of PCR products, possibly by preventing reannealing of the strands. Formamide, glycerol, DMSO, Tween 20, and NP-40 are suggested as remedy for difficulties in amplification of GC-rich templates [23] as well as betaine at 1 M [24], 1.3 M [25], 2 M [26] or at 1 M in combination with DMSO [27, 28]. DMSO disrupts DNA base pairing without affecting fidelity [29], while betaine has been reported to affect the base pair composition dependence of DNA strand composition [30].

Trehalose [31], protein BSA and gelatin stabilize the DNA polymerase during thermal cycling. Nonionic detergents Tween 20 and NP-40 might have a beneficial effect in this respect, as they are added to Taq DNA polymerase purification protocols for this reason [32]. Tween 20 and NP-40 alone or in combination with DMSO have been also reported to improve specificity and raise yield of PCR in general [33] and also neutralize negative effects of SDS [34].

Relatively new PCR enhancers are nanoparticles from gold (AuNPs) [35, 36], titanium dioxide (TiO_2) [37] and graphene oxide (GO) or reduced graphene oxide (rGO) [38]. The mode of action of nanoparticles has not been elucidated in detail. Finally, attempts have been made to combine several enhancers in an attempt to find synergistic positive effects [39–41].

Future developments of yeast colony PCR protocols should separate the effects of DNA release and PCR inhibition, and how these effects vary with variables such as culture medium, age and growth phase of the cultures, and then systematically apply the relevant conditions based on the results for other direct PCR protocols.

5   Notes

1. The choice of PCR primers can be important. PCR primers should be specific for both vector and insert, as false positive detection may arise by using PCR primers that are specific only for the insert or vector. The explanation for this surprising phenomena is that DNA from the ligation mixture may adsorb onto the surface of the cells and serve as a PCR template, masking the absence of the correct DNA construct inside the cells [42].

2. Each colony must be transferred to both culture medium and PCR reaction. Since many clones are usually screened, keeping track of PCR tubes and clones may be a logistical issue. We have found that stabbing a gridded LB plate with the tip and leaving it there is a good way of keeping track of the picked clones.

3. Many published protocols rely on the use of sterile toothpicks for transferring clones to the PCR tubes. It should be noted that toothpicks have been associated with PCR inhibitors [43]. If this is a concern, sterile pipet tips can be used instead.

4. We have found that toothpicks may absorb some of the PCR mix if cells are added into the mastermix. Pipette tips used in the same way may also remove some PCR mix by capillary action. We deposit the cells above the surface of the PCR mix in the tube and vortex the tubes prior to PCR. This has the added benefit of not allowing interaction between PCR mix and template prior to PCR.

5. We usually keep an open petri dish with a suitable solid selective yeast medium nearby to preserve the clones. The petri dish is gridded with 8×8–10×10 squares using a marker pen or by placing on a printable petri dish grid [44]. The toothpicks can be left standing in the agar as a help to keep track of processed clones.

6. This protocol is sensitive to the amount of yeast cells in the PCR tube. During the set-up of this protocol, it is useful to use a PCR test case. We use primers 19_D-DFR1 (5′ GAC TCA GAC AGG TTG AAA AGA AGA C 3′) and 18_A-DFR1 (5′ CAA AGG TTT GGT TTT CAG TTA AGA A 3′) to amplify a 1288 bp PCR product from the *DFR1* locus in *S. cerevisiae* using a program consisting of initial denaturation for 4 min at 94 °C, followed by 30 cycles of 94 °C for 30s, 50 °C for 30s, and 72 °C for 45 s, and a final extension at 72 °C for 5 min. This PCR reaction is very robust and any yeast colony PCR protocol should do it with success.

7. The original protocol also states that a 10 min incubation at room temperature can be performed instead of the incubation at 70 °C, but the high temperature should inactivate nucleases that can potentially degrade DNA. This might be an issue since DNA and cell debris are present together until the last step.

8. It is important to remove as much as possible of the ethanol in **step 9** of protocol in Subheading 3.2, as ethanol could be a PCR inhibitor. The tubes can be incubated in a 37 °C heat block for 3–5 min in order to evaporate traces of ethanol.

Acknowledgment

This work was supported by the strategic programme UID/BIA/04050/2013 (POCI-01-0145-FEDER-007569) funded by national funds through the FCT I.P. and by the ERDF through the COMPETE2020—Programa Operacional Competitividade e Internacionalização (POCI).

References

1. Güssow D, Clackson T (1989) Direct clone characterization from plaques and colonies by the polymerase chain reaction. Nucleic Acids Res 17:4000

2. Gibson DG, Benders GA, Axelrod KC, Zaveri J, Algire MA, Moodie M, Montague MG, Venter JC, Smith HO, Hutchison CA (2008) One-step assembly in yeast of 25 overlapping DNA fragments to form a complete synthetic mycoplasma genitalium genome. Proc Natl Acad Sci U S A 105:20404–20409

3. Pereira F, Azevedo F, Parachin NS, Hahn-Hägerdal B, Gorwa-Grauslund MF, Johansson B (2016) Yeast pathway kit: a method for metabolic pathway assembly with automatically simulated executable documentation. ACS Synth Biol 5:386–394

4. Flávio Azevedo Humberto Pereira (2016) Online yeast colony PCR protocols. In: Public Github Gist. https://gist.github.com/BjornF Johansson/490ca933976d286cbaef37a07df4 86b8. Accessed 1 Jul 2016

5. Sathe GM, O'Brien S, McLaughlin MM, Watson F, Livi GP (1991) Use of polymerase chain reaction for rapid detection of gene insertions in whole yeast cells. Nucleic Acids Res 19:4775

6. Ling M, Merante F, Robinson BH (1995) A rapid and reliable DNA preparation method for screening a large number of yeast clones by polymerase chain reaction. Nucleic Acids Res 23:4924–4925

7. Wang H, Kohalmi SE, Cutler AJ (1996) An improved method for polymerase chain reaction using whole yeast cells. Anal Biochem 237:145–146

8. Bourke MT, Scherczinger CA, Ladd C, Lee HC (1999) NaOH treatment to neutralize inhibitors of Taq polymerase. J Forensic Sci 44:1046–1050

9. Akada R, Murakane T, Nishizawa Y (2000) DNA extraction method for screening yeast clones by PCR. BioTechniques 28:668–670. 672, 674

10. Linke B, Schröder K, Arter J, Gasperazzo T, Woehlecke H, Ehwald R (2010) Extraction of nucleic acids from yeast cells and plant tissues using ethanol as medium for sample preservation and cell disruption. BioTechniques 49:655–657

11. Lõoke M, Kristjuhan K, Kristjuhan A (2011) Extraction of genomic DNA from yeasts for PCR-based applications. BioTechniques 50:325–328

12. Rossen L, Nørskov P, Holmstrøm K, Rasmussen OF (1992) Inhibition of PCR by components of food samples, microbial diagnostic assays and DNA-extraction solutions. Int J Food Microbiol 17:37–45

13. Gietz RD, Schiestl RH (2007) High-efficiency yeast transformation using the LiAc/SS carrier DNA/PEG method. Nat Protoc 2:31–34

14. Pham TA, Kawai S, Murata K (2011) Visualization of the synergistic effect of lithium acetate and single-stranded carrier DNA on Saccharomyces cerevisiae transformation. Curr Genet 57:233–239

15. Harju S, Fedosyuk H, Peterson KR (2004) Rapid isolation of yeast genomic DNA: bust n' grab. BMC Biotechnol 4:8

16. Blount BA, Driessen MRM, Ellis T (2016) GC Preps: fast and easy extraction of stable yeast genomic DNA. Sci Rep 6:26863

17. Walsh PS, Metzger DA, Higuchi R (1991) Chelex 100 as a medium for simple extraction of DNA for PCR-based typing from forensic material. BioTechniques 10:506–513

18. Kermekchiev MB, Kirilova LI, Vail EE, Barnes WM (2009) Mutants of Taq DNA polymerase resistant to PCR inhibitors allow DNA amplification from whole blood and crude soil samples. Nucleic Acids Res 37:e40

19. Wang Y, Prosen DE, Mei L, Sullivan JC, Finney M, Vander Horn PB (2004) A novel strategy to engineer DNA polymerases for enhanced processivity and improved performance in vitro. Nucleic Acids Res 32:1197–1207

20. Ghadessy FJ, Ong JL, Holliger P (2001) Directed evolution of polymerase function by compartmentalized self-replication. Proc Natl Acad Sci U S A 98:4552–4557

21. Baar C, d'Abbadie M, Vaisman A, Arana ME, Hofreiter M, Woodgate R, Kunkel TA, Holliger P (2011) Molecular breeding of polymerases for resistance to environmental inhibitors. Nucleic Acids Res 39:e51

22. Winship PR (1989) An improved method for directly sequencing PCR amplified material using dimethyl sulphoxide. Nucleic Acids Res 17:1266

23. Varadaraj K, Skinner DM (1994) Denaturants or cosolvents improve the specificity of PCR amplification of a G + C-rich DNA using genetically engineered DNA polymerases. Gene 140:1–5

24. Henke W, Herdel K, Jung K, Schnorr D, Loening SA (1997) Betaine improves the PCR amplification of GC-rich DNA sequences. Nucleic Acids Res 25:3957–3958

25. Hengen PN (1997) Optimizing multiplex and LA-PCR with betaine. Trends Biochem Sci 22:225–226

26. Mytelka DS, Chamberlin MJ (1996) Analysis and suppression of DNA polymerase pauses associated with a trinucleotide consensus. Nucleic Acids Res 24:2774–2781

27. Frackman S, Kobs G, Simpson D, Storts D et al (1998) Betaine and DMSO: enhancing agents for PCR. Promega Notes 65:27–29

28. Kang J, Lee MS, Gorenstein DG (2005) The enhancement of PCR amplification of a random sequence DNA library by DMSO and betaine: application to in vitro combinatorial selection of aptamers. J Biochem Biophys Methods 64:147–151

29. Hardjasa A, Ling M, Ma K, Yu H (2010) Investigating the effects of DMSO on PCR fidelity using a restriction digest-based method. J Exp Microbiol Immunol 14:161–164

30. Rees WA, Yager TD, Korte J, von Hippel PH (1993) Betaine can eliminate the base pair composition dependence of DNA melting. Biochemistry 32:137–144

31. Spiess A-N, Mueller N, Ivell R (2004) Trehalose is a potent PCR enhancer: lowering of DNA melting temperature and thermal stabilization of taq polymerase by the disaccharide trehalose. Clin Chem 50:1256–1259

32. Desai UJ, Pfaffle PK (1995) Single-step purification of a thermostable DNA polymerase expressed in *Escherichia coli*. BioTechniques 19(780–782):784

33. Bachmann B, Lüke W, Hunsmann G (1990) Improvement of PCR amplified DNA sequencing with the aid of detergents. Nucleic Acids Res 18:1309

34. Wilson IG (1997) Inhibition and facilitation of nucleic acid amplification. Appl Environ Microbiol 63:3741–3751

35. Li H, Huang J, Lv J, An H, Zhang X, Zhang Z, Fan C, Hu J (2005) Nanoparticle PCR: nanogold-assisted PCR with enhanced specificity. Angew Chem Int Ed 44:5100–5103

36. Yang W, Li X, Sun J, Shao Z (2013) Enhanced PCR amplification of GC-rich DNA templates by gold nanoparticles. ACS Appl Mater Interfaces 5:11520–11524

37. Khaliq RA, Sonawane PJ, Sasi BK, Sahu BS, Pradeep T, Das SK, Mahapatra NR (2010) Enhancement in the efficiency of polymerase chain reaction by TiO 2 nanoparticles: crucial role of enhanced thermal conductivity. Nanotechnology 21:255704

38. Jia J, Sun L, Hu N, Huang G, Weng J (2012) Graphene enhances the specificity of the polymerase chain reaction. Small 8:2011–2015

39. Musso M, Bocciardi R, Parodi S, Ravazzolo R, Ceccherini I (2006) Betaine, dimethyl sulfoxide, and 7-deaza-dGTP, a powerful mixture for amplification of GC-rich DNA sequences. J Mol Diagn 8:544–550

40. Ralser M, Querfurth R, Warnatz H-J, Lehrach H, Yaspo M-L, Krobitsch S (2006) An efficient and economic enhancer mix for PCR. Biochem Biophys Res Commun 347:747–751

41. Zhang Z, Kermekchiev MB, Barnes WM (2010) Direct DNA amplification from crude clinical samples using a PCR enhancer cocktail and novel mutants of Taq. J Mol Diagn 12:152–161

42. Dallas-Yang Q, Jiang G, Sladek FM (1998) Avoiding false positives in colony PCR. BioTechniques 24:580–582

43. Lee AB, Cooper TA (1995) Improved direct PCR screen for bacterial colonies: wooden toothpicks inhibit PCR amplification. BioTechniques 18:225–226

44. Colony Immunoblotting Assay for Detection of Bacterial Cell-surface or Extracellular Proteins —BIO-PROTOCOL. http://www.bio-protocol.org/e888. Accessed 26 Jul 2016

CRENAME, A Molecular Microbiology Method Enabling Multiparametric Assessment of Potable/Drinking Water

Luc Bissonnette, Andrée F. Maheux, and Michel G. Bergeron

Abstract

The microbial assessment of potable/drinking water is done to ensure that the resource is free of fecal contamination indicators or waterborne pathogens. Culture-based methods for verifying the microbial safety are limited in the sense that a standard volume of water is generally tested for only one indicator (family) or pathogen.

In this work, we describe a membrane filtration-based molecular microbiology method, CRENAME (Concentration Recovery Extraction of Nucleic Acids and Molecular Enrichment), exploiting molecular enrichment by whole genome amplification (WGA) to yield, in less than 4 h, a nucleic acid preparation which can be repetitively tested by real-time PCR for example, to provide multiparametric presence/absence tests (1 colony forming unit or microbial particle per standard volume of 100-1000 mL) for bacterial or protozoan parasite cells or particles susceptible to contaminate potable/drinking water.

Key words Drinking water analysis, Molecular microbiology, Multiparametric detection, Microbial assessment, Fecal contamination indicators, Waterborne pathogens, CRENAME, Membrane filtration, Whole genome amplification

1 Introduction

Drinking water (or potable water used for drinking) is essential to sustain the life of human populations and this resource is generally considered safe when it is free of fecal contamination indicators (FCI) such as *Escherichia coli*, enterococci, and total coliforms, or of potential microbial human pathogens such as *Clostridium perfringens*, *Cryptosporidium parvum*, *Cryptosporidium hominis*, and *Giardia intestinalis*. In analytical terms, "free" means that the tested water contains less than one colony forming unit (CFU) or microbial particle (MP; spore or [oo]cyst for example) of the sought indicator or pathogen per standard volume of 100 mL (general water testing), 250 mL (bottled water in Europe), or 1–10 L (for *Cryptosporidium* and/or *Giardia*).

Lucília Domingues (ed.), *PCR: Methods and Protocols*, Methods in Molecular Biology, vol. 1620,
DOI 10.1007/978-1-4939-7060-5_9, © Springer Science+Business Media LLC 2017

In 1999, after the Interlaken workshop on molecular technologies for safe drinking water, the Organisation for Economic Co-operation and Development (OECD) published a Health Policy Brief [1] which provided us with guidance for the inception of an innovative water molecular microbiology programme for supporting the development and operation of a mobile environmental laboratory, Atlantis [2].

In light of the fact that FCI seldom fail to provide an adequate evaluation of the risk posed by waterborne pathogens such as viruses (e.g., *Norovirus*), bacteria (e.g., *Vibrio cholerae*), and protozoan parasites (*Cryptosporidium* spp., *G. intestinalis*, etc.) [1, 3–6], we believe that a better approach would be to enable the concomitant or consecutive testing of a standard water sample for FCI and selected waterborne pathogens. However, this is not possible with current culture-based methodologies since the standard volume (e.g., 100 mL) is generally completely used for interrogating a single (and quite specific) culture medium.

The CRENAME (Concentration Recovery Extraction of Nucleic Acids and Molecular Enrichment) molecular microbiology method [7] was designed in part to circumvent this limitation for bacterial and parasitic waterborne pathogens (containing a DNA genome) and to address a series of principles or elements concerning (drinking) water microbial assessment and methodological stumbling blocks of current methods or strategies:

1. FCI also generally fail to provide an adequate evaluation of the risk posed by fastidious or non culturable waterborne pathogens, or of those capable of entering into a viable but non culturable (VBNC) state [8].

2. Methods currently approved for the microbial assessment of a water sample prevent the retrospective analysis of a standard sample for confirmation of a positive result or for the search for another FCI or pathogen.

3. Some tests approved for testing drinking water are based on the presence/absence principle meaning that detection is positive when 1 or more CFU or MP is present in the standard volume tested. In most instances, the mere detection of an FCI or of a specific pathogen would be sufficient to call for a public health response or, at least, retesting of the water sample or source for confirmation of contamination.

4. Membrane filtration is a widely used method for primary concentration of water samples prior to incubation of the filter on a selective and/or chromogenic solid medium used to grow and enumerate bacterial FCI or pathogens. As contaminated drinking water may transiently contain a wide range of FCI and pathogens, the predictive value of culture-based methods

is limited by their selectivity; for example, testing for *E. coli* when enterococci are present and vice versa. In such conditions, an alternate sample preparation procedure enabling the efficient recovery and detection of low numbers of different MPs immobilized on or within a cellulose ester filtration membrane would be warranted.

5. Molecular enrichment by whole genome amplification (WGA; [9–11]) is methodologically similar to microbiological (culture) enrichment, as done in many culture-based methods to increase the probability of detection of microbial targets even those identified in 1° [12–16]. WGA is an unbiased amplification process catalyzed by phage φ29 DNA polymerase capable of providing large amounts of DNA from a small number of genomic DNA molecules. The product of WGA is a DNA solution that can be stored and repetitively tested by molecular biology analytical processes.

6. Molecular tools such as multiplex polymerase chain reaction (PCR), real-time PCR (rtPCR), microarrays, and biosensors can be exploited for the rapid and specific multiparametric detection of microbial genetic targets.

7. The method can be tailored to operate within the physical, methodological, and ecological limitations of a mobile infrastructure [2].

As most DNA purification procedures are not efficient enough to recover DNA at low concentration, WGA is used to increase the amount of FCI and waterborne pathogens DNA at levels detectable by rtPCR; in fact, 1 or 2 μL of WGA-enriched sample can be used to provide a presence/absence detection signal by rtPCR. Thus, CRENAME is essentially a molecular microbiology procedure permitting the efficient recovery of as low a one microbial particle per standard volume of drinking water (100–1000 mL) following filtration of a standard volume on a 0.45 μm membrane made of mixed cellulose esters; CRENAME can provide a WGA-enriched sample in less than 4 h [7].

To the best of our knowledge, filtration membrane dissolution was pioneered by Aldom and Chagla for the detection of *Cryptosporidium* oocysts [17], but we perfected it for specific applications in drinking water microbial assessment and, during the development of CRENAME, demonstrating that detection of bacteria was feasible opened a wider range of applications in water molecular microbiology. Briefly, after filtration of a standard water volume on a conventional manifold, the microbial particle(s) is(are) liberated from the filter matrix by disintegrating said matrix in the presence of methanol, and then completing its dissolution in the presence of acetone. Methanol and acetone are solvents known to exert a fixative effect on cells [18, 19]. During filter disintegration, the suspension of particles, containing those from the sample and bacterial spores spiked to serve as process control, is put in presence of glass beads which, along with filter membrane remnants, contribute to the confinement of very low numbers of microbial particles. At one point during the process, the contents of the tested standard water volume (100–1000 mL) are concentrated in a volume of ca. 2 mL in a microcentrifuge tube. After high-speed centrifugation and removal of the supernatant, the concentrated microbial particles are contained in a volume of 30–40 μL (concentration factor varying from 3000 to 30,000 times depending on standard volume tested). The glass beads will then be instrumental for extracting the nucleic acids of concentrated particles by mechanical lysis prior to 3-h molecular enrichment step by WGA. The WGA-enriched nucleic acid product can be interrogated multiple times by PCR or rtPCR [7].

Coupled to rtPCR, CRENAME can provide a presence/absence result for the detection of a bacterial FCI or parasite (oo) cyst is less than 5 h. However, when considering the workflow of CRENAME, the method cannot yet be applied for the detection of waterborne viruses [20, 21], since 0.45 μm filtration membranes would be suboptimal for their recovery and their concentration by low-speed centrifugation is merely impossible [22]. Finally, WGA is not suitable for amplifying RNA targets.

In the context of drinking water microbial assessment, CRENAME was shown to be equivalent to the USEPA Method 1604 on MI agar for the detection of *E. coli*/Shigella (1.8 CFU/100 mL; [7]). We also reported that CRENAME-rtPCR enabled the detection of enterococci (4.5 CFU/100 mL; [23]) and *C. perfringens* spores (1 CFU/100 mL; [24]) in potable/drinking water, including well groundwater. Furthermore and as proof of the concept of multiparametric microbial assessment of drinking water, we have exploited CRENAME-rtPCR to simultaneously test well water samples for the presence of *E. coli*, *Enterococcus* spp., *Enterococcus faecalis/faecium*, and *Bacillus atrophaeus* subsp. *globigii* [25]. Recently, CRENAME has also been used to develop a presence/absence test for the (more heterogeneous) group of total coliforms in drinking water [26]. Finally, to provide a procedure faster than Method 1623 for the assessment of source water for the presence of protozoan parasites under the Long Term 2 Enhanced Surface Water Treatment Rule (LT2ESWTR; [27]), CRENAME-rtPCR has the potential to detect as few as two *C. parvum* oocysts and one *G. intestinalis* cyst per liter of potable water in less than 5 h [unpublished work].

2 Materials

The CRENAME procedure can be separated in three distinct steps after membrane filtration: (1) dissolution of filtration membrane, and confinement and concentration of microbial particles, (2) nucleic acid extraction, and (3) molecular enrichment by WGA. This section describes the equipment, materials, and reagents required for performing the procedure.

2.1 Water Sample Membrane Filtration

1. Potable/drinking water sample. Volume from 100 to 1000 mL (*see* **Note 1**).

2. Internal process control. *B. atrophaeus* subsp. *globigii* CCRI-9827 (equivalent to strain NRS1221 A; also named *B. globigii*) [28] (*see* **Note 2**).

3. Tri-headed standard filtration manifold, UV sterilization apparatus, and vacuum pump.

4. Sterile (sterilizable) tweezers or forceps.

5. Sterile non-gridded GN-6 Metricel membrane filters (47 mm diameter, 0.45 μm pore size; PALL Corporation) (*see* **Notes 3 and 4**).

6. Reverse osmosis-purified water (RO water; resistivity of 18 MΩ-cm min at 25 °C).

7. Sterile RO water is autoclaved for 30 min (121 °C, 15 lbs./in^2).

8. Sterile 15-mL polypropylene tube.

2.2 Dissolution of Filtration Membrane, and Confinement and Concentration of Microbial Particles

1. HPLC-grade methanol.

2. Histological-grade acetone.

3. 2-mL screw-cap microcentrifuge tube.

4. Sterile acid-washed glass beads (150–212 μm and 710–1180 μm; Sigma-Aldrich) (*see* **Note 5**).

5. Sterile 1× TE buffer: 10 mM Tris–HCl, 1 mM EDTA, pH 8.0.
   1× TE buffer is made by diluting 10 mL of 1 M Tris pH 8.0 and 2 mL of 500 mM EDTA pH 8.0 with 800 mL of RO water. The pH is verified and adjusted with 1 or 6 N HCl, and the volume is completed to 1000 mL. The solution is filtered through 0.1 μm disposable units directly into 500-mL sterile glass bottles, sterilized by autoclaving for 30 min (121 °C, 15 lbs./in^2), and dispensed in individual tubes as aliquots of 1 mL before storage at −20 °C.
   1 M Tris pH 8.0. 12.1 g of Tris base is dissolved into 80 mL of RO water. pH is adjusted to 8.0 with 6 N HCl and the volume is completed to 100 mL. The stock solution is dispensed as 10 mL aliquots and stored at −20 °C.

500 mM EDTA pH 8.0. 18.6 g EDTA.2H$_2$O disodium salt, and 2 g NaOH are dissolved in 80 mL of RO water. pH is adjusted to 8.0 with 6 N HCl or 6 N NaOH and the volume is completed to 100 mL. The stock solution is dispensed as 10 mL aliquots and stored at −20 °C.

6. Vortex mixer.

7. Clinical centrifuge.

8. Microcentrifuge.

9. 200- and 1000-μL micropipettors.

2.3 Nucleic Acid Extraction and Molecular Enrichment by WGA

1. Illustra GenomiPhi™ V2 DNA Amplification Kit (GE Healthcare Life Sciences) (*see* **Note 6**).

2. Micropipettors.

3. Thermomixer or thermal cycler.

3 Methods

3.1 Water Sample Membrane Filtration

1. Prior to filtration, *B. globigii* spores used as internal process control are added to water samples at approximately 60 per 100 mL.

2. For filtration of the potable/drinking water sample (100–1000 mL) spiked with *B. globigii* spores, a GN-6 membrane is aseptically deposited on the filtration platform of a tri-headed manifold before assembly of the UV-sterilized funnel with filtration head. The water sample is dispensed in the funnel and vacuum is applied to force the sample through the filtration membrane. The funnel is rinsed with approximately 25–30 mL of sterile RO water and the filtration is completed.

3. After filtration, the funnel is removed and the filtration membrane is aseptically and quickly transferred to a clean 15-mL polypropylene tube (*see* **Note 7**) (Fig. 1).

3.2 Dissolution of Filtration Membrane, and Confinement and Concentration of Microbial Particles

1. To initiate the disintegration of the membrane, 8.5 mL of HPLC-grade methanol is added to the 15-mL polypropylene tube and incubated for 10 s, before being vigorously agitated on a vortex mixer for approximately 10 s (*see* **Note 8**). The reaction tube and its contents are then centrifuged for 3 min at 2100 × *g*, and the supernatant is removed and discarded (*see* **Note 9**). To complete the dissolution of the filtration membrane materials, 0.75 mL of histological-grade acetone is added to the pellet and membrane remnants are dissolved by vigorous agitation on a vortex mixer (*see* **Note 10**).

2. The clear acetone solution resulting from the dissolution of the filtration membrane and containing the microbial particles and

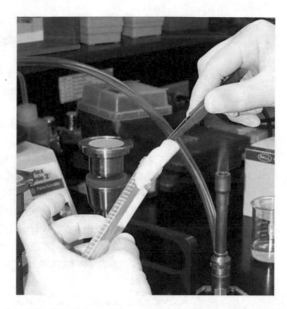

Fig. 1 Transfer of the filtration membrane into a 15-mL polypropylene tube. While still wet, the filtration membrane is rolled over sterile tweezers or forceps to facilitate insertion in the polypropylene tube. The top portion of the filter (where most particles must be immobilized) facing inward the tube to minimize physical or physicochemical interactions of particles with the plastic surface

B. globigii spores (process control) is transferred to a 2-mL microcentrifuge tube containing a mixture of sterile acid-washed glass beads. The 2-mL tube is centrifuged for 3 min at 15,800 × *g* to confine the MPs and glass beads, and the supernatant is carefully removed and discarded (*see* **Notes 11** and **12**).

3. The 15-mL polypropylene tube is briefly rinsed with 1.0 mL of histological-grade acetone and the resulting mixture is transferred to the 2-mL microtube containing glass beads. The tube is then centrifuged for 3 min at 15,800 × *g*, and the supernatant is removed and discarded (*see* **Notes 11** and **12**).

4. The resulting pellet is washed, to reduce the organic solvent content, by carefully adding 1.0 mL of sterile 1× TE buffer to the 2-mL tube, without perturbing the pellet, followed by a centrifugation of 3 min at 15,800 × *g*. After centrifugation, the supernatant is removed and discarded without touching the pellet containing confined MPs and glass beads (*see* **Notes 11** and **12**). The residual volume, estimated to 30–40 μL, includes a dead volume of approximately 15–25 μL.

3.3 Nucleic Acid Extraction

1. Forty (40) microliters of Illustra GenomiPhi™ V2 sample buffer is added to the slurry containing microbial particles and glass beads. The microbial particles and (fixed) cells are mechanically lysed by the glass beads during 5 min of vigorous mixing, at room temperature, at maximum speed on a vortex mixer.

2. The reaction tube containing the crude nucleic acid extract is incubated 3 min at 95 °C for (a) melting genomic DNA, (b) inactivating nucleases and other inhibitors, and (c) enabling the hybridization of random hexamers to DNA. The mixture is quenched on ice for a minimum of 3 min.

3.4 Molecular Enrichment by WGA

1. A reaction mixture composed of 45 μL of GenomiPhi™ reaction buffer and 4 μL of Phi29 (φ29) DNA polymerase is added to the crude nucleic acid extract, mixed gently by finger tapping, before a brief spun in a microcentrifuge. At this point, the reaction volume is estimated to 105–115 μL.

2. Molecular enrichment by WGA is performed by incubating the reaction mixture for 3 h at 30 °C (*see* **Note 6**). The enzymatic reaction is arrested by an incubation of 10 min at 65 °C and WGA-amplified materials can be stored at −80 °C for future analysis.

3. For detection of microbial targets (FCI or waterborne pathogens, and *B. globigii* process control) by rtPCR amplification, 1 μL of WGA-amplified products is generally used as template.

4   Notes

1. CRENAME was primarily developed for testing nonchlorinated potable water used for drinking such as commercially available ozonated spring water and raw well groundwater. For the time being, the method has not been reliable for testing water which may contain a wide range of inhibitors of WGA and/or PCR.

2. For the assessment of drinking water by molecular microbiology, water samples are spiked with *B. globigii* spores prior to membrane filtration. In bioanalytical terms, these spores serve as internal process control for (1) the recovery of microbial particles, (2) the extraction of nucleic acids, and (3) rtPCR inhibition, and their DNA is detected by rtPCR. The spores are prepared as described by Picard et al. [28].

3. The use of non-gridded membrane filters is essential since the materials (inks) used to make the grid are insoluble in methanol and acetone, and inhibit PCR amplification.

4. In our hands, only the membranes from PALL Corporation (cat. no. 66265) provided good recovery of microbial particles.

5. Commercially available glass beads should be acid-washed and rinsed with sterile RO water to remove inhibitors of the WGA reaction, before being dispensed in 2-mL microtubes (approximately 10 mg per tube). Sterilization by autoclaving is done under a "Liquid Cycle" for 30 min (121 °C, 15 lbs./in²), to minimize static clinging of the beads to the plastic surfaces of the microtubes.

6. During the development of CRENAME, several WGA kits have been tested but, so far, the Illustra GenomiPhi™ V2 DNA Amplification Kit was found to be better adapted for the procedure. In an ideal world, the components of this kit should be reworked to improve the workflow of the method and minimize wasting costly reagents.

7. To better control the disintegration of the filtration membrane, it must be transferred while still moist to the polypropylene tube and not allowed to dry as this would accelerate disintegration in the presence of methanol. For easier insertion in the tube, the membrane is rolled over the sterilized tweezers or forceps, with the top of the filter facing inward the tube (*see* Fig. 1).

8. For adequate recovery of microbial particles, the disintegration of the membrane must be arrested when filter fragments of ca. 5 × 5 mm are obtained. These fragments are easier to pellet along with MPs and this facilitates the removal of the supernatant (mostly methanol).

9. If the separation between the pellet and supernatant is not sufficient, another centrifugation is immediately repeated.

10. The user would be advised to keep approximately 2–3 mm of the methanol supernatant over the pellet before adding acetone, to prevent disturbing the pellet and losing MPs.

11. To avoid the loss of confined particles, the supernatant must be carefully removed by aspiration using a 200-µL micropipettor. Up to 5 mm of supernatant can be left over the pellet at this point in the procedure without significant loss of performance.

12. During the confinement of MPs in the presence of glass beads, it is crucial NOT TO VORTEX OR MIX, to avoid initiating the nucleic acid extraction process which would result in the loss of liberated nucleic acids. Nucleic acid extraction will be done during the onset of WGA molecular enrichment.

Acknowledgments

During her doctoral thesis, Andrée F. Maheux held a scholarship from the Nasivvik Center for Inuit Health and Changing Environment (Canadian Institutes of Health Research) and the work was supported in part by the Canada Foundation for Innovation.

Author contributions: L. Bissonnette has written the chapter in collaboration with A.F. Maheux, while M.G. Bergeron was responsible for its revision.

Conflicts of Interests: The authors declare having no conflict of interest.

References

1. OECD (Organisation for Economic Co-operation and Development) (1999) Health Policy brief – molecular technologies for safe drinking water: results from the interlaken workshop, Switzerland. http://www.oecd.org/health/biotech/2097510.pdf. Accessed 16 Jun 2016

2. Maheux AF, Bissonnette L, Huppé V et al (2016) The requirements and challenges of a mobile laboratory for onsite water microbiology assessment. Water Pract Technol 11:198–209. doi:10.2166/wpt.2016.024

3. Costán-Longares A, Montemayor M, Payán A et al (2008) Microbial indicators and pathogens: removal, relationships and predictive capabilities in water reclamation facilities. Water Res 42:4439–4448

4. USEPA (United States Environmental Protection Agency) (2008) Literature review of molecular methods for simultaneous detection of pathogens in water. EPA/600/R-07/128, Environmental Technology Council, United States Environmental Protection Agency, Cincinnati, OH, p 139. http://nepis.epa.gov/Adobe/PDF/P1008BJ3.pdf. Accessed 16 Jun 2016

5. Payment P, Locas A (2011) Pathogens in water: value and limits of correlation with microbial indicators. Ground Water 49:4–11

6. Wu J, Long SC, Das D et al (2011) Are microbial indicators and pathogens correlated? A statistical analysis of 40 years of research. J Water Health 9:265–278

7. Maheux AF, Bissonnette L, Boissinot M et al (2011) Rapid concentration and molecular enrichment approach for sensitive detection of Escherichia coli/Shigella in potable water samples. Appl Environ Microbiol 77:6199–6207

8. Li L, Mendis N, Trigui H, Oliver JD et al (2014) The importance of the viable but non-culturable state in human bacterial pathogens. Front Microbiol 5:258

9. Lasken RS, Egholm M (2003) Whole genome amplification: abundant supplies of DNA from precious samples or clinical specimens. Trends Biotechnol 21:531–535

10. Lovmar L, Syvänen A-C (2006) Multiple displacement amplification to create a long-lasting source of DNA for genetic studies. Hum Mutat 27:603–614

11. Binga EK, Lasken RS, Neufeld JD (2008) Something from (almost) nothing: the impact of multiple displacement amplification on microbial ecology. ISME J 2:233–241

12. Scheusner DL, Busta FF, Speck ML (1971) Inhibition of injured Escherichia coli by several selective agents. Appl Microbiol 21:46–49

13. Feng PC, Hartman PA (1982) Fluorogenic assays for immediate confirmation of Escherichia coli. Appl Environ Microbiol 43:1320–1329

14. Frahm E, Obst U (2003) Application of the fluorogenic probe technique (TaqMan PCR) to the detection of Enterococcus spp. and Escherichia coli in water samples. J Microbiol Meth 52:123–131

15. Dwivedi HP, Jaykus L-E (2011) Detection of pathogens in foods: the current state-of-the art and future directions. Crit Rev Microbiol 37:40–63

16. Sontakke S, Cadenas MB, Maggi RG et al (2009) Use of broad range 16S rDNA PCR in clinical microbiology. J Microbiol Meth 76:217–225

17. Aldom JE, Chagla AH (1995) Recovery of Cryptosporidium oocysts from water by membrane filter dissolution method. Lett Appl Microbiol 20:186–187

18. Fukatsu T (1999) Acetone preservation: a practical technique for molecular analysis. Mol Ecol 8:1935–1945

19. Mangels JI, Cox ME, Lindberg LH (1984) Methanol fixation. An alternative to heat fixation of smears before staining. Diagn Microbiol Infect Dis 2:129–137

20. Ganesh A, Lin J (2013) Waterborne human pathogenic viruses of public health concern. Int J Environ Health Res 23:544–564

21. Gibson KE (2013) Viral pathogens in water: occurrence, public health impact, and available control strategies. Curr Opin Virol 4:50–57

22. Ikner LA, Gerba CP, Bright KR (2012) Concentration and recovery of viruses from water: a comprehensive review. Food Environ Virol 4:41–67

23. Maheux AF, Bissonnette L, Boissinot M et al (2011) Method for rapid and sensitive detection of Enterococcus sp. and Enterococcus faecalis/faecium cells in potable water samples. Water Res 45:2342–2354

24. Maheux AF, Bérubé È, Boudreau DK et al (2013) Abilities of the mCP agar method and CRENAME alpha toxin-specific real-time PCR assay to detect Clostridium perfringens spores in drinking water. Appl Environ Microbiol 79:7654–7661

25. Maheux AF, Huppé V, Bissonnette L et al (2012) Comparative analysis of classical and molecular microbiology methods for the detec-

tion of *Escherichia coli* and *Enterococcus* spp. in well water. J Environ Monitor 14:2983–2989

26. Maheux AF, Boudreau DK, Bisson M-A et al (2014) Molecular method for detection of total coliforms in drinking water samples. Appl Environ Microbiol 80:4074–4084

27. USEPA (U.S. Environmental Protection Agency) (2006) National primary drinking water regulations: long term 2 enhanced surface water treatment rule; final rule. Fed Register 71:654–786

28. Picard FJ, Gagnon M, Bernier MR et al (2009) Internal control for nucleic acid testing based on the use of purified *Bacillus atrophaeus* subsp. *globigii* spores. J Clin Microbiol 47:751–575

Chapter 10

Multiplex Detection of Food-Borne Pathogens

Germán Villamizar-Rodríguez and Felipe Lombó

Abstract

Detection of food-borne pathogens is traditionally carried out by plating out techniques in selective or differential media using Petri agar dishes or other culture-dependent methods, usually designed for each pathogen to be detected. These classical methods are time and personnel consuming and also may last for up to 5 days in the case of final confirmation of some specific pathogens.

Here we describe a method for fast multiplex detection of nine food-borne pathogens (all species usually required under most countrylegislations) by means of a single multiplex PCR reaction coupled to a capillary electrophoresis detection, in just 2–2.5 h and with a minimum cost of around 2 € per sample and nine pathogens. This method saves consumables and personnel time and allows a faster detection of any possible contaminated food batches at industrial level, therefore helping to prevent future food-borne outbreaks at clinical level.

Key words Food-borne pathogens, Universal preenrichment medium, Multiplex PCR

1 Introduction

Food safety and quality is a continuous challenge for scientists, food industry and public authorities. Some pathogens such as *Bacillus cereus*, *Campylobacter jejuni*, *Clostridium perfringens*, *Cronobacter sakazakii*, *Escherichia coli* O157:H7, *Enterobacteriaceae* Family, *Listeria monocytogenes*, *Salmonella* spp. and *Staphylococcus aureus* represent a serious health risk for consumers, and therefore, its detection and/or quantification is a mandatory issue in many countries. Diverse national or international agencies such as the European Food Safety Agency (EFSA) or the U.S. Food and Drug Administration (FDA) establish the corresponding food safety regulations, which are associated to the specific food matrix and pathogens involved. To accomplish these regulatory frames represents a big challenge for the different parts involved in the food manufacturing and distribution chain. The conventional detection method relies in plating out and growing those pathogens which may be potentially present in a given food matrix. This standard

Lucília Domingues (ed.), *PCR: Methods and Protocols*, Methods in Molecular Biology, vol. 1620,
DOI 10.1007/978-1-4939-7060-5_10, © Springer Science+Business Media LLC 2017

method has been useful for many years, but it involves a high level of laboratory work and time needed to obtain confirmation of pathogens presence. In some cases, up to 96 h may be necessary in order to obtain confirmation, delaying the implementation of the right actions to stop or contain an outbreak.

Molecular methods, such as PCR, allow fast detection of food-borne pathogens, avoiding extended time frames and intense human laboratory work. Also, these molecular methods produce results in less time than conventional culture ones. PCR detection of pathogens, allows an important reduction in time analysis, generating results in just 21–24 h, including an initial pre-enrichment step. This allows a faster response in case of an outbreak, but it also reduces storage costs for the food industry.

PCR detection methods can be improved, by using more than one pair of primers in the same PCR reaction, allowing the detection of more than one DNA target simultaneously in multiplex mode. Multiplex PCR has been used previously for food-borne pathogens detection [1]. This technique is able to detect up to nine different targets at the same time [2], showing a high potential as the routine method in different food safety analyses.

Also, a non-selective pre-enrichment is a necessary step in pathogen detection. This step is used to promote growth of putative pathogens which are present in low numbers or which may have been affected by manufacturing, transport or storage conditions, causing a drop in the numbers of viable cells, which could not be detected in a given assay. In order to simplify our detection method and to save time and reagents, a universal pre-enrichment medium, named GVUM [2] has been developed in this work. This GVUM has been designed to allow growth of most important food-borne pathogens, such as the nine pathogens described above.

The multiplex PCR method described here coupled to capillary electrophoresis, represents an advantage in food-borne pathogen detection, because of the possibility of performing amplification of nine different targets in the same reaction, with a high level of accuracy and resolution, reducing time and sample processing.

2 Materials

2.1 Primers Design

1. gDNA database such as GenBank (NCBI).
2. Blastn/Blastp online tools [3].
3. Primer3 online primer design tool [4].
4. FastPCR Software [5].

2.2 Multiplex Primer Optimization and PCR Runs

1. Amplitaq Gold 360 Master Mix (Applied Biosystems) (*see* **Note 1**).

2. Multiplex Primers set for nine selected pathogens [2].

3. PCR 200 μL tubes, Stripes or 96 well plates.

4. Double distilled, Nuclease Free or Milli-Q H₂O.

5. Thermal Cycler.

6. Vortex mixer.

7. 0.5–10, 10–100 and 100–1000 μL Micropipettes with their corresponding tips.

8. gDNA extracted from pure cultures of each microorganism.

2.3 Sampling and Pre-Enrichment

1. Sampling recipients.

2. Spatula, spoons and forks.

3. Laboratory scale.

4. Homogenizer: Stomacher® 80 (Seward) or similar.

5. Whirl-Pak® (Nasco) 384 mL capacity bags, with filter membrane, for stomacher.

6. Universal pre-enrichment medium (GVUM [2]): 20 g/L Buffered Peptone Water, 50 mL/L Hemolyzed and Defibrinated horse blood, 20 g/L Mannitol, 4 mL/L *Campylobacter* Growth Supplement OXOID SR0232, Distilled Water.

7. Laboratory Incubator.

2.4 Genomic DNA Extraction

1. Qiagen DNeasy Blood & Tissue Kit (Qiagen), for gDNA extracted from pure cultures (standard gDNA).

2. PrepMan® Ultra Reagent (Thermo Fisher Scientific) for gDNA extracted from food samples.

3. 1.5 mL Eppendorf® Safe-Lock Microcentrifuge tubes.

4. Hot water bath or thermal block.

5. Microcentrifuge.

6. 0.5–10, 10–100 and 100–1000 μL Micropipettes with tips.

2.5 Agarose Gel Electrophoresis of DNA

1. Electrophoresis Buffer Tris-Borate-EDTA (TBE) 1×: 108 g/L Tris–HCl, 20 mL 0.5 M EDTA pH = 8, 55 g/L Boric Acid, Distilled Water.

2. High purity Molecular Grade Agarose (low electroendosmosis grade).

3. Laboratory scale.

4. Microwave oven.

5. Fisherbrand™ Glass media bottles.

6. RedSafe™ 20,000× DNA dye (Chembio) or DNA-Dye non tox (PanReac AppliChem).

7. 0.5–10, 10–100 and 100–1000 µL Micropipettes with tips.

8. Electrophoresis Kit (trays, combs, chamber and power source).

9. UV Transilluminator/Gel documentation system.

2.6 Capillary Electrophoresis of DNA (Lab-On-Chip)

1. 2100 Bioanalyzer® (Agilent).

2. DNA 1000® Chips (Agilent).

3. DNA 1000® Reagents.

4. Desktop or laptop PC with the 2100 Expert software (Agilent) installed.

5. Vortex with Chip adaptor (part of the 2100 Bioanalyzer Kit).

3 Methods

3.1 Primer Design

1. For primer design, it is necessary to choose DNA targets which must be specific for each pathogen. For this purpose, genomes of each pathogen must be acquired from DNA databases such as GenBank. In the selection process, it is necessary to avoid genes with high level of variation or those related with virulence/resistance features which may be present only in some strains of a given species. The selected target gene must show a well preserved sequence or must be common for all strains of the pathogen of interest.

2. Once the gene sequence target has been established, different software available on-line or in desktop versions can be used for designing the PCR primers. Primer3 [4] and Fast PCR [5] are useful tools which allow to fix final size of amplicons. This is a necessary step in order to obtain amplicons with different sizes which will be visualized later. These softwares also allow the addition of GC-clamps at the end of the selected sequences. Fast PCR allows primer evaluation too, in order to avoid primer-dimer and cross amplification.

3. At the moment of ordering the synthesis of the designed oligonucleotides, you may order them as lyophilized product, or already dissolved in some buffers. We usually prefer the first option, doing the stock solution in TE 1× in our laboratory, on a final concentration of 100 µM for each primer.

3.2 Multiplex Primer Optimization

1. Test individual performance of each primer pair with the corresponding gDNA target, using a concentration of 1 µM for each primer and fixing the Tm following the data obtained from the software during the primer design process.

2. Adjust the concentration of each primer at a given Tm range, testing this Tm range starting 2 °C under the theoretical Tm and increasing 2 °C in each trial.

3. Test the following primer concentrations: 0.4, 0.6, 0.8 and 1 µM for each oligonucleotide.

4. Perform the PCR and choose those two best combinations of concentration and Tm temperature, which have produced the highest yield of PCR products.

5. Choose a second pair of primers and test them in a duplex reaction, evaluating the amplification result with those two combinations of Tm and concentrations.

6. Choose the better combination from that previous duplex reaction and proceed to add another primer pair, testing it again for a combination of Tm/concentration.

7. Sequentially add new primer pairs until the desired amplifications are obtained, using all primers and gDNA targets in the same reaction (*see* **Note 2**).

3.3 Food Matrices Sampling

1. Food sampling procedures are well regulated by the food safety agencies. A model for sampling can be consulted online in the laboratory manuals of FDA [6]. For example, for the genus *Salmonella*, FDA establishes three categories of food which have to be considered during the sampling. For each category, I, II and III, the number of analytical units (25 g each) are 60, 30 and 15, correspondingly.

2. Samples can be processed immediately or can be kept at 4 °C until use.

3.4 Pre-Enrichment

1. Mix all ingredients of the GVUM medium in a beaker; add enough distilled water to guarantee good mixing. Transfer the mix to a graduate cylinder or to another volumetric device. Complete the desired volume with distilled water. Then sterilize it autoclaving at 121 °C during 15 min.

2. In a sterile cabinet, weigh 25 g of the corresponding food sample in the homogenizer bag, using a sterile fork or spatula to transfer the food matrix. Assure that the food portion reaches the bottom of the bag avoiding the introduction of the fork or spatula. If the sample has been retained in the middle, use the bag walls to displace it to the right place.

3. Add to the bag the GVUM medium previously prepared, directly from the sterile bottle until reaching 100 g of total weigh, considering the sample and the pre-enrichment medium.

4. Seal the bag, folding from the opened size and using a clip.

5. Insert the bag in the homogenizer (Stomacher or similar). Homogenize for 1 min for soft samples or 2 min for difficult food samples.

6. Remove the bag from the homogenizer and place it in the incubator for 24 h at the corresponding temperature (37–42 °C).

3.5 Genomic DNA Extraction

1. Remove the bag from the incubator and open it carefully, in order to avoid splashing.

2. Using a sterile serological pipette, collect aliquots of 1 mL from each bag and place them in 1.5 mL centrifuge tube. Remember to use one pipette to each bag. Discard the pipette and the bag properly.

3. Centrifuge the tubes at $13,523 \times g$, for 1 min.

4. Discard supernatant and add 180 µL of PrepMan® Ultra Reagent.

5. Mix well by vortex or pipetting.

6. Place the tubes in a thermal block or water bath at 100 °C for 10 min. Sometimes caps could open during the process, be careful with hot splash.

7. Once this incubation time is finished, remove the tubes and let cool to room temperature.

8. Centrifuge at 12,000 rpm for 2 min.

9. Collect 100 µL of the supernatant and place it in a new sterile 1.5 mL tube.

10. Keep at 4 °C until use (*see* **Note 3**).

3.6 PCR Run

1. Prepare the Master Mix, following the instructions of the manufacturer and considering optimized concentration for each primer pair. Remember to consider at least one or two reactions more than the number of samples to test, in order to balance possible pipetting errors during preparation.

2. Place 1–2 µL of DNA samples from Subheading 3.5 in the PCR tubes, strips or plates.

3. Add the corresponding volume of master mix.

4. Place the tubes/strips/plates in the thermal cycle and run the PCR amplification program (*see* **Note 4**).

5. Remove the samples and keep at 4 °C, waiting for the next step.

3.7 Agarose Gel Electrophoresis of DNA

1. 5% Agarose preparation: add 5 g agarose in 100 mL of TBE buffer and mix well.

2. Heat in the microwave until dissolution (~ 3 min), being careful for explosive boiling.

3. Assure that all agarose is dissolved. If not, place it again in the microwave and heat for 30 s more.

4. Let cool down for few minutes and then add 5 µL/100 mL of DNA dye (*see* **Note 5**).

5. Mix well avoiding excessive formation of bubbles.

6. Prepare an electrophoresis tray and place the agarose in it.

7. Distribute uniformly the agarose and introduce the comb.

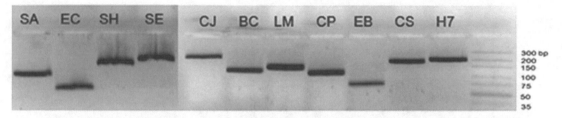

Fig. 1 Agarose gel electrophoresis showing the nine individual PCR amplicons generated on total DNA obtained from spiked food matrix (*red minced meat*) with the different pathogens of this study. The right bar shows the molecular weight of the corresponding PCR amplicons. BC: *Bacillus cereus*, CJ: *Campylobacter jejuni*, CP: *Clostridium perfringens*, CS: *Cronobacter sakazakii*, EB: *Enterobacteriaceae*, EC: *E. coli*, LM: *Listeria monocytogenes*, SA: *Staphylococcus aureus*, SE: *Salmonella* spp.

8. Once the gel is solid, remove the comb, taking care to not break the gel.

9. Add the corresponding volume of loading buffer in each sample tube (containing PCR product) mixing well by pipetting (*see* **Note 6**).

10. Place each sample in the well, remembering to include a DNA Ladder lane (*see* **Note 7**).

11. Close the electrophoresis chamber, programming the power unit for the right voltage and time (*see* **Note 8**).

12. Start electrophoresis run.

13. Once the run has been finished, remove the gel and place it in a UV lamp (*see* **Note 9**). A typical result is shown in Fig. 1.

3.8 Capillary Electrophoresis of DNA (Lab-On-Chip)

Traditional capillary electrophoresis methods have been progressively substituted by novel miniaturized devices, which allow cleaner, accurate and fast electrophoresis runs, such as those based on Lab-on-chip (LOC) technology approach. LOC devices have a web of micro-channel and wells, which have been impressed on an inner glass, silica or polymeric surface of the cartridge. Pressure or electrokynetic forces, induces the displacement of the sample through the channel, allowing the separation of the different fragment DNA or molecule, which will be detected by fluorescence in the analyzer [7].

The following instructions are described for LOC DNA-chips and kits of the 2100 Bioanalyzer System, other devices or systems may require different procedures:

1. Allow Gel-Dye mix to equilibrate to room temperature for 30 min.

2. Place the DNA 1000 chip in the priming station, positioning the syringe plunger at 1 mL.

3. Once the Gel-Dye temperature has been equilibrated, pipette 9 μL in the right well (marked with a rounded "G"), taking care to not introduce bubbles in the chip.

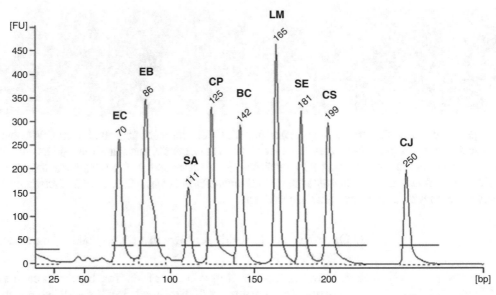

Fig. 2 Capillary electropherogram showing the amplification of the nine pathogen gene targets after the multiplex PCR. BC: *Bacillus cereus*, CJ: *Campylobacter jejuni*, CP: *Clostridium perfringens*, CS: *Cronobacter sakazakii*, EB: *Enterobacteriaceae*, EC: *E. coli*, LM: *Listeria monocytogenes*, SA: *Staphylococcus aureus*, SE: *Salmonella* spp.

4. Close the priming station and press the plunger until hearing a "click".

5. Wait 60 s and then release the plunger.

6. Before 5 s, pull back gently the plunger up to 1 mL position and open the priming station (*see* **Note 10**).

7. Pipette now 9 µL in each one of the another two wells marked with a squared "G".

8. Load 5 µL of Gel-Dye Mix in the 12 sample wells, taking care to not leave any well empty or introduce bubbles in the well.

9. Put the chip in the vortex adapter and mix for 1 min at $28.97 \times g$, avoiding splashes (*see* **Note 11**).

10. Place the chip on the 2100 Bioanalyzer, close the lid and check that the chip has been recognized by the software.

11. Start the run.

12. Results can be viewed, analyzed and edited in the data window of the 2100 Expert software (Fig. 2).

4 Notes

1. Based in our experience, we recommend the Amplitaq Gold 360 Master Mix (Applied Biosystems). However, another commercial PCR Master Mix or separately DNA polymerase can be used, but in any case a previous optimization can be performed.

2. It is possible to use new concentration of primers during the optimization process. Sometimes, initial concentration parameters must be increased or decreased for certain primers in order to enhance the amplification of some targets which have produced weak amplicons during the optimization steps.

3. Applied Biosystems recommends keeping the samples at 4 °C until a maximum lapse of 1 month. If the samples have to be processed after 1 month, keep it at −20 °C.

4. The PCR program varies according to the DNA polymerase used and to the length of the targets and expected amplicons. Using Amplitaq Gold 360 Master Mix and the primers described in [2], conditions are the following: (1) Initial denaturation step (95 °C, 180 s, 1 cycle); (2) 35 cycles of denaturation Step (95 °C, 30 s), plus annealing step (56 °C, 60 s), plus elongation Step (72 °C, 90 s) (3) Final elongation step (72 °C, 420 s).

5. Use of non-toxic DNA Dye, instead ethidium bromide, is strongly recommended. Different commercial alternatives are available; we use routinely RedSafe™ 20,000× DNA dye (Chembio) or DNA-Dye non tox (PanReac AppliChem) obtaining good results. The first one must be added to the agarose and the second one to the loading buffer.

6. Loading Buffers (LB) are composed by a mix of dyes such as xylene-cyanol or cresol red, and density agents such as sucrose, ficoll or glycerol. These allow the correct loading of DNA in the well and show the relative position of the DNA in the gel during electrophoresis. The amount that should be added is dependent on the total volume of the sample and concentration of the Ladder. If 6× LB is used, add 1 part of LB to 5 part of sample. If 10× LB is used, add 1 of LB to 9 parts of sample.

7. DNA Ladder is needed to check the size of PCR products in agarose electrophoresis. Ladders are available in different DNA size ranges. Choose the one that fits in the size of your PCR products.

8. General rule says increasing voltage, the time for electrophoresis decrease. So, high voltage in low concentrated agarose gels, produce a "smile" effect in bands or in worst cases, gel or sample degradation. 90–120 V is an average voltage for most agarose gels. Adjust time to length of the gel, using loading buffer colored front as an indicator. Perform some runs, in order to adjust the time and voltage that produce best results with your samples.

9. Remove the gel from the chamber, using a paper napkin to absorb the excess of TBE buffer. Place the gel in the transilluminator carefully, taking care to not break it. Wear eye and skin protection during the operation of UV lamps.

10. Some resistance pulling the plunger is natural. If the plunger doesn't move by itself during the first seconds or it is impossible to pull, probably the chip is defective or the gel charge has been incorrectly performed. In this case, discard the chip and repeat the process.

11. Starting with at $11.31–20.12 \times g$, increasing up to 2400 is recommended to avoid splashes.

Acknowledgments

Authors wish to thank Government of Principality of Asturias-FICYT for financial support (Grants PEST08-17, IE09-257C1 and IE09-257C2).

References

1. Asakura M, Samosornsuk W, Hinenoya A, Misawa N, Nishimura K, Matsuhisa A et al (2008) Development of a cytolethal distending toxin (cdt) gene-based species-specific multiplex PCR assay for the detection and identification of Campylobacter jejuni, Campylobacter coli and Campylobacter fetus. FEMS Immunol Med Microbiol 52:260–266

2. Villamizar-Rodríguez G, Fernández J, Marín L, Muñiz J, González I, Lombó F (2015) Multiplex detection of nine food-borne pathogens by mPCR and capillary electrophoresis after using a universal pre-enrichment medium. Front Microbiol 6:1194

3. Altschul SF, Gish W, Miller W, Myers EW, Lipman DJ (1997) Basic local alignment search tool. J Mol Biol 215:403–410

4. Rozen S (1998) Primer3. http://bioinfo.ut.ee/primer3-0.4.0/primer3/

5. Kalendar R, Lee D, Schulman AH (2011) Java web tools for in silico PCR and oligonucleotide assembly and analysis. Genomics 98:137–144

6. Food and Drug Administration-FDA (2016) Bacteriological analytical manual (BAM). https://www.fda.gov/Food/FoodScience Research/LaboratoryMethods/ucm2006949.htm

7. Yoon JY, Kim B (2012) Lab-on-a-Chip pathogen sensors for food safety. Sensors (Basel) 12(8):10713–10741

Chapter 11

Fast Real-Time PCR for the Detection of Crustacean Allergens in Foods

Francisco J. Santaclara and Montserrat Espiñeira

Abstract

Crustaceans are one of the most common allergens causing severe food reaction. Hypersensitivity reactions associated with seafood intake are one of the most common food allergies in adults. Crustaceans including shrimps, prawns, crabs, lobster, and crayfish are a common cause of anaphylaxis or hypersensitivity, with shrimps and crabs being the most common causes of allergy. Symptoms occur most often when food or cooking water are ingested.

These food allergens are a health problem, and they have become very important; as evidenced by the existence of several regulations that establish that labeling must be present regarding these allergens to warn consumers.

The methodology herein exposed allows the detection of crustaceans in any type of product, including those where very aggressive treatments of temperature and pressure are used during the manufacturing process.

The main features of this method are its high sensitivity and specificity, and reduced analysis time of real-time PCR (40 min). This assay is a potential tool in issues related to the labeling of products and food security to protect the allergic consumer.

Key words Crustacean, Detection, Allergen, Fast real-time PCR, LNA probe

1 Introduction

An allergen is a substance that can induce a hypersensitivity or allergic reaction in susceptible people who have previously been in contact with this substance. Food allergens can act at very low doses (traces) and potentially cause fatal reactions depending on factors such as the individual's tolerance or health, because there is no curative therapy for these allergic diseases; and the most effective way to prevent allergic reactions is to remove the sensitive components that trigger adverse effect from the diet.

For this reason, consumers should have all the necessary information on food composition on the label, to make a choice that suits their needs. Changes in labeling legislation have allowed a significant improvement in the labeling of ingredients with allergenic capacity.

Lucília Domingues (ed.), *PCR: Methods and Protocols*, Methods in Molecular Biology, vol. 1620,
DOI 10.1007/978-1-4939-7060-5_11, © Springer Science+Business Media LLC 2017

The traces are minimal amounts of a substance or ingredient, which are included in a food. The presence of trace amounts of allergenic substances may occur due to cross-contamination or because they were included in one of the raw materials used in the manufacture of a product. When there is a demonstrable risk for the presence of these substances in trace quantities they must be declared as a precaution, with a preventive label containing a warning message.

Allergens can also be present in some products as a result of common practice in the food industry and do not appear in the label. This is mainly due to cross-contamination between raw materials, production lines, or equipment, caused by insufficient cleaning of them. For this reason, it acquires special attention conducting surface controls to detect these traces of allergens in plants where food production is done.

Given all the issues previously highlighted, detection methods are needed for determining the presence or absence of allergenic substances in trace amounts, because small amounts are sufficient to produce allergic reactions in sensitized individuals.

In this chapter a protocol based on the real-time PCR TaqMan assay using specific set primers/probe is described, which allows the detection of crustaceans in food in trace amounts, and responds to the needs of the food industry, allowing to indicate the presence/absence of crustaceans in food.

2 Materials, Equipment, and Reagents

2.1 Materials

1. Sterile microtubes of 1.5 and 2 mL.
2. MicroAmp® Optical 96-Well Reaction Plate (Applied Biosystems).
3. Optical adhesive films.
4. Tubes of 50 mL.
5. Foil.
6. Sterile bags.

2.2 Equipments

1. Thermomixer (Eppendorf).
2. Vortemp.
3. Refrigerated centrifuges for microtubes.
4. Micropipettes of variable volume and their respective tips (0.1–2.5 µL; 0.5–10 µL; 10–100 µL; 100–1000 µL; 1–10 mL).
5. Laminar flow hoods or PCR cabinet.
6. Hot plate.
7. Water purification system (Milli-Q, Millipore).
8. Tweezers.

9. Spectrophotometer (NanoDrop).

10. Vortex mixer.

11. Tissue homogenizer.

12. Autoclave.

13. Balances.

14. Freezer.

15. Centrifuge for PCR plates.

16. Maxwell® 16 MDx Research Instrument.

17. ViiA 7 Real-time PCR System (Applied Biosystems).

2.3 Reagents

1. Maxwell® 16 FFS Nucleic Acid Extraction Kit X9431 (Promega).

2. CTAB (cetyl trimethyl ammonium bromide). For 1 L of Maxwell CTAB buffer: 20 g of CTAB, 81.9 g of NaCl, 100 mL of 1 M Tris hydrochloride solution, 40 mL of 0.5 M EDTA pH 8.0. Dissolve and bring to 1 L (to facilitate the dissolution of the components, preheat on a hot plate with stirring at 65 °C). Autoclave for sterilizing the solution.

3. NaCl.

4. 1 M Tris.

5. 0.5 M EDTA pH 8.

6. TaqMan® Universal PCR Master Mix (Applied Biosystems).

7. Oligonucleotide primers (Table 1).

8. FAM™-LNA labeled oligonucleotide probe (Table 1).

3 Methods

3.1 DNA Extraction

1. Prepare CTAB buffer as described in the Subheading 2.3.2, use within 3 months, store capped in the refrigerator at 4 °C. Preheat the CTAB solution at 65 °C before use to dissolve the precipitate particles.

Table 1
Primers and probe used in this protocol for the detection of crustaceans in food with the protocol herein described

Oligonucleotide designation	Sequence (5′–3′)	Molecular marker	Amplicon size (bp)	Reference
Forward primer	TAA AGT CTG GCC TGC CCA	16S rRNA	205	[1]
Reverse primer	GCT TTA TAG GGT CTT ATC GT			
Probe	6-FAM-TGC TAC CTT IGC ACG GTC A-BHQ1(LNA)			[2]

2. Perform DNA extraction using a kit for automated isolation. The example given is for the Maxwell® 16 FFS Nucleic Acid Extraction Kit X9431. The procedure with sample varies depending on the type of sample, which can be classified into two groups:

3. For the DNA extraction from the sample, continue as described:

Sample Type 1: fish, meat, and processed meat products (raw and cooked, as for instance cured products, tinned food, convenience food, hamburgers, doner kebab). Proceed as described below:

(a) Weigh 50 mg of sample material into a 1.5 mL tube.

(b) Add 600 µL CTAB buffer, 30 µL of proteinase K (20 mg/mL), and 2 µL of RNase A solution.

(c) Vortex vigorously and incubate for at least 60 min at 60 °C in a thermo shaker (1400 rpm).

(d) Centrifuge for 10 min at $\geq 14{,}000 \times g$.

Sample Type 2: food, feed, seed (e.g., rapeseed, soybean, corn, rice, chocolate bars, milk powder). Proceed as described below:

(a) Grind sample material and weigh 100 mg (rapeseed, soybean) or 200 mg (corn) into a 2 mL tube.

(b) Add 1 mL CTAB buffer, 40 µL of proteinase K (20 mg/mL) and optionally 20 µL of RNase A solution.

(c) Vortex vigorously and incubate for 90 min at 65 °C.

(d) Centrifuge for 10 min at $16{,}000 \times g$ (optional: centrifuge supernatant again).

(a) Place the cartridge to be used into the Maxwell LEV cartridge rack and remove the seal.

(b) Pipet 300 µL of sample supernatant and 300 µL of Lysis Buffer into well 1 of the Maxwell cartridge (Fig. 1).

(c) Place one of the supplied elution tubes into the sample rack (position 2) and add 100 µL of the supplied elution buffer (Fig. 1).

(d) Place the plunger in the position 3 of the cartridge (Fig. 1).

(e) Select LEV configuration of the Maxwell and select method: RUN → DNA → Blood → Start run.

(f) After the extraction your sample is ready-to-use for downstream applications.

4. The DNA obtained from the samples must be quantified with a spectrophotometer using the absorbance at 260 and 280 nm, and the concentration must be normalized to use in the downstream steps (*see* **Note 1**).

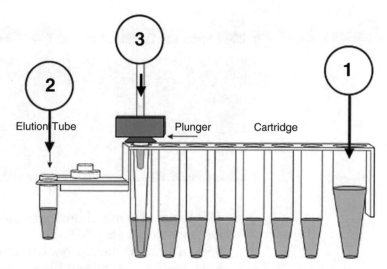

Fig. 1 Schematics of a Maxwell cartridge. *Circled numbers* refer to positions described in the text

Table 2
Composition of PCR reactions

Mix components	Mix 1	Mix N
TaqMan® Universal PCR Master Mix (2×)	10 μL	10 × N μL
Forward primer 10 μM	1.8 μL	1.8 × N μL
Reverse primer 10 μM	1.8 μL	1.8 × N μL
Probe 10 μM	0.5 μL	0.5 × N μL
Molecular biology grade water	to 20 μL	to 20 × N μL

3.2   Real-Time PCR

While the PCR reactions are being set up, precautions should be taken at all stages to avoid contamination of samples and reagents (*see* **Note 2**).

1. Mix well all reagents by inverting the tubes a number of times or for small volumes by flicking the tube and spin briefly (~5 s) in a centrifuge.

2. Prepare a PCR master mix in 1.5 mL microcentrifuge tubes, following the proportions indicated in the Table 2 to maximize uniformity and minimize labor when working with multiple tubes or plates. The final volume of each master mix depends on the number of reactions required for each real-time PCR. Master mixes can be made in excess of what is actually needed to account for pipetting variations, i.e., prepare 10% excess of master mix. The probe must be handled with extreme caution (*see* **Note 3**).

Table 3
PCR amplification profile of the real-time PCR system herein described

Step	Time (s)	Temperature (°C)
Hold	20	95
PCR (×40 cycles)		
Denaturation	1	95
Annealing	20	60

3. Mix the master mix thoroughly and dispense equal aliquots into each thin-walled PCR tube. Set up the real-time PCR in a 96-well plate, including nontemplate (NTC), negative controls, positive controls and the test samples, each at least in duplicate (*see* **Note 4**).

4. Add DNA template (125 ng of DNA per sample to each well) to each reaction tube as needed (be sure to include a negative control that does not receive any template). When pipetting is finished, seal the plate with optical adhesive film.

5. Centrifuge the reaction plate at low speed for 1 min and then load the plate onto a ViiA 7 real-time PCR System.

3.3 Program and Run Applied Biosystems ViiA 7 System

1. Operate the Applied Biosystems ViiA 7 System and SDS (Sequence Detection System) software according to the manufacturer's recommendations. Set reaction volume to 20 μL, and the amplification program as described in Table 3 (unless other parameters are established in the internal methodological validation). Use FAM/LNA as detector/quencher.

2. Briefly, launch the SDS software and open a new plate template window to denote well locations on the 96-well plate for the controls, and testing samples. Save the template window with the recorded data as a SDS run file.

3. Select the "Connect" button, and then load the 96-well plate into the instrument. Select the "Start" button to begin the actual run.

4. Run the PCR reaction to determine Ct (cycle threshold) values for each sample.

3.4 Data Collection and Analysis

After the run is completed, select baseline, and place the threshold line at the exponential phase of amplification. Remove the plate from the instrument and save the results of the run. When the analysis button is selected in the SDS software, the results will be analyzed automatically if the standards and testing sample information were recorded as noted in the previous section. The Ct values obtained should be thoroughly reviewed (*see* **Note 5**) and interpretation of the results must be made in relation to data obtained in the internal validation performed in each laboratory for a given detection limit (*see* **Notes 6–8**), allowing to determine whether a sample contains or does not contain crustaceans in its composition.

4   Notes

1. The isolation and purification of DNA from the sample constitutes one of the most important steps of any real-time PCR protocol. The quality and integrity of both are essential to carry out an experiment successfully. One way to measure these characteristics of the DNA is using a spectrophotometer, measuring the absorbance at A260 and A280 nanometers. The ratio A260/A280 should be higher than 1.8. When lower values are obtained, this may indicate the presence of impurities or contaminants in the DNA solution as proteins or phenol, and which may affect the assay. Higher values than 2 may be due to the presence of RNA in the DNA solution. If the DNA extracted is not enough (less than 200 ng) or the purity is not adequate, DNA extraction should be repeated.

2. The following precautions should be taken to avoid contamination. It is paramount to work in at least three separate areas. One area should be designated just for DNA extraction; other for the preparation of the master mix; and other area designated just for adding the template. This is crucial to avoid contamination of the real-time PCR reagents with aerosols containing the template. Pipette sets must be dedicated for use in these areas and must not be moved between areas. Use only dedicated pipettes for preparing master mixes and another set for adding template DNA. Use only filter tips and sterile plasticware. Wear gloves and change them frequently.

3. TaqMan probe and probe dilution: probes are usually received as a 100 pmol/μL stock solution. This stock solution can be diluted with molecular biology grade water to achieve a working solution concentration. Probe working solutions should be aliquoted in volumes that are sufficient for one time use. These aliquots should be stored at −20 °C and protected from light. In addition, repeated freeze–thaw cycles of probe solutions should be avoided.

4. Negative reaction control (NTC) produces amplification. The DNA may be contaminated. This may be due to amplicon contamination in the RT reagents or, in some cases, contamination in the DNA preparation. It is often best to simply repeat the real-time PCR reaction using new reagents (except DNA extraction). If the problem disappears, one can proceed. If the problem persists, it is necessary to make a new DNA extraction.

The positive control is a tube of DNA extracted from food matrix reference material contaminated with crustaceans at the detection limit of the technique. To determine this, serial dilutions of crustacean DNA and DNA from different species of seafood can be prepared to a final amount of 125 ng [2].

5. Test samples should be analyzed in duplicate or triplicate according to individual laboratory statistical requirements. Excessive variability between Ct of duplicate or triplicate samples: if the samples display large variability between duplicates and triplicates there may be a problem in the preparation of the master mix or a pipetting error. Check that the master mix was prepared properly and well mixed, and that the micropipette is calibrated and is dispensing the proper amount of liquid reproducibly.

6. Before applying these techniques in a particular laboratory, internal validations should be made taking into account the following points:

    Thermocyclers: As previously mentioned, each thermocycler has its own characteristics that may require an adjustment in the working parameters (temperature or reaction times, as well as temperature ramp rate).

    DNA polymerase enzyme: It is also necessary to make a validation if the polymerase employed is different from the herein described, since polymerases from different commercial dealers may vary in amplification efficiency, and consequently may affect the optimum temperature and the specificity of the technique, generating variations in observed cycle threshold and amplification profile.

    Substances included in the validation: In the phase of methodological design and development of the detection system a great number of substances is included for evaluating the specificity of the technique, and so validate the correct working. However, each laboratory should perform an internal validation including all the matrixes used in their production chain, since these could interfere with the results, giving for instance false positives.

7. The detection limit should be calculated on the basis of the validation performed on each laboratory, as it will depend on the equipment and technical personnel. Similarly, the cut off or Ct, which separates a positive sample of a negative will vary depending on the specific detection limit of each particular laboratory.

8. To consider a test as valid, it must have a correct reading of the controls (the positive control in the detection limit must be positive, and the NTC and negative controls should not have amplification). Moreover, the amplification curve should have an exponential appearance, and the threshold cycle must be less than the Ct determined in the methodological validation.

References

1. Brzezinski JL (2005) Detection of crustacean DNA and species identification using a PCR–restriction fragment length polymorphism method. J Food Prot 68(9):1866–1873

2. Herrero B, Vieites JM, Espiñeira M (2012) Fast real-time PCR for the detection of crustacean allergen in foods. J Agric Food Chem 60(8):1893–1897

Chapter 12

Fast Real-Time PCR Method for Detection of Soy in Foods

Montserrat Espiñeira and Francisco J. Santaclara

Abstract

Soy is used as an additive in the manufacturing of diverse products, because of their ability of emulsification, water and fat absorption, contributing to the consistency of food products. Moreover, soy is recognized as a potential allergen, so its presence should be indicated in all the food products.

These issues highlight the need for techniques that allow the detection of soy in foods. This work describes a real-time PCR method for the detection of soy in a wide range of foodstuffs. The main features of this technique are its reliability and sensitivity, allowing the detection of trace amounts of soybean in processed products. TaqMan real-time PCR is one of the simplest and fastest molecular biology techniques, with a high potential for automation. Therefore, it is one of the techniques most used for screening a variety of substances.

The methodology herein described is of great value in issues regarding the presence of soy protein in processed products, especially in verifying labeling and security regulations to protect consumer's rights.

Key words Detection, Allergen, Emulsifier, Adulteration, Authenticity, Additives

1 Introduction

In the food industry, many of the additives used in the manufacture of diverse products are vegetable protein, being soybeans one of the most used and referenced [1, 2]. Among the properties of soy stand out the emulsification capabilities and the water and fat absorption, contributing to more consistent food products. In the case of meat products and seafood, the water absorption and retention properties of textured soy proteins can be used to bind moisture in fish or meat blocks and reduce fat and jelly deposits liberated during the manufacturing process, obtaining a firmer final product. Increased water retention results in end products with a higher weight; which means that soy protein replaces raw material, usually with a higher cost. This makes the food products more cost-effective; there is a considerable decrease in production costs to the benefit of the manufacturers. In addition to the fraudulent use of soy, the other negative side is its allergenicity, as it is recognized as a potential allergen by the *Codex Alimentarius* [3]. Therefore, the

Lucília Domingues (ed.), *PCR: Methods and Protocols*, Methods in Molecular Biology, vol. 1620,
DOI 10.1007/978-1-4939-7060-5_12, © Springer Science+Business Media LLC 2017

detection of this food additive in food products focuses on ensuring food quality and safety, as the adulteration of food not only causes harm to consumers, but also unfair competition between industries in the food sector.

Due to all the issues outlined above, a methodology to detect soy protein in food products is necessary to ensure food safety, consumer protection and, at the same time, to guarantee fair competition between producers. This protocol is based on the real-time PCR TaqMan assay using specific set primers/probe developed by Espiñeira et al. [4].

2 Materials, Equipment, Reagents, and Preparations

2.1 Materials

1. Sterile microtubes of 1.5 and 2 mL.
2. MicroAmp® Optical 96-Well Reaction Plate (Applied Biosystems).
3. Optical adhesive films.
4. Tubes of 50 mL.
5. Foil.
6. Sterile bags.

2.2 Equipments

1. 1.Thermomixer (Eppendorf).
2. Vortemp.
3. Refrigerated centrifuges for microtubes.
4. Variable volume micropipettes of and their respective tips (0.1–2.5 μL; 0.5–10 μL; 10–100 μL; 100–1000 μL; 1–10 mL).
5. Laminar flow hoods or PCR cabinet.
6. Hot plate.
7. Water purification system (Milli-Q, Millipore).
8. Tweezers.
9. Spectrophotometer (NanoDrop).
10. Vortex mixer.
11. Tissue homogenizer.
12. Autoclave.
13. Balances.
14. Freezer.
15. Centrifuge for PCR plates.
16. ViiA 7 Real-time PCR System (Applied Biosystems).

2.3 Reagents

1. 1 M Tris (Tris hydroxymethyl aminomethane), pH 8.0. For 1 L of final volume, add the following amounts: 121.1 g of

Tris, 700 mL of Milli-Q H_2O. Dissolve the Tris and bring to 900 mL. Adjust the pH to 8.0 with concentrated HCl (will need ~50 mL). Bring to 1 L.

2. 5 M NaCl. For 1 L of final volume, add the following amounts: 292.2 g of NaCl, 700 mL of Milli-Q H_2O. Dissolve and bring to 1 L.

3. NaOH pellets.

4. 0.5 M EDTA (ethylenediaminetetraacetic Acid), pH 8.0. For 1 L of final volume, add the following amounts: 186.12 g of EDTA, 750 mL of Milli-Q H_2O. Add about 20 g of NaOH pellets. Slowly add more NaOH until pH is 8.0, EDTA will not dissolve until the pH is near 8.0. Bring to 1 L.

5. CTAB (cetyl trimethyl ammonium bromide). For 1 L of CTAB buffer: 100 mL of 1 M Tris, pH 8.0, 280 mL of 5 M NaCl, 40 mL of 0.5 M EDTA, 20 g of CTAB. Add Milli-Q H_2O to 1 L. To facilitate the dissolution of the components, preheat on a hot plate with stirring at 65 °C.

6. RNAasA.

7. Polyvinylpyrrolidone mol. Weight 40,000 (PVP-40). β-mercaptoethanol.

8. Proteinase k.

9. Chloroform–isoamyl alcohol (24:1).

10. 7.5 M Ammonium Acetate. For a final volume of 250 mL, add the following amount: 144.5 g of ammonium acetate and bring to volume with Milli-Q H_2O.

11. Isopropanol (2-propanol).

12. 70 and 95% ethanol.

13. TaqMan® Universal PCR Master Mix (Applied Biosystems).

14. Oligonucleotide primers (Table 1).

15. FAM™-TAMRA™-labeled oligonucleotide probe (Table 1).

Table 1
Primers and probe used in this protocol for the soy detection with the protocol herein described

Oligonucleotide designation		Sequence (5′-3′)	Molecular marker	Amplicon size (bp)	References
Forward primer	LEC IF	CTT CTT TCT CGC ACC AAT	Lectin gene	100	[5]
Reverse primer	LEC IR	CTC AAC AGC GAC GAC TTG			
Probe	LE-Probe	6-FAM-CAC ATG CAG GTT ATC TTG GTC-TAMRA			[1]

3 Methods

3.1 DNA Extraction

1. Prepare CTAB solution as described in the materials section (Subheading 2.3, **item 5**), use within 3 months, store capped in the refrigerator at 4 °C. Prior to use, add to 100 mL of CTAB, 4 g polyvinylpyrrolidone MW 40,000 (PVP-40) and 500 µL β-mercaptoethanol and stir to dissolve before starting extractions. Preheat the CTAB solution at 65 °C before use to dissolve the precipitate particles.

2. Weigh out 10 g of triturated and homogenized sample that will be analyzed.

3. Add 10 mL of preheated CTAB buffer and 100 µL proteinase K (10 mg/mL).

4. Transfer the solution to a 50-mL tube. *Optional:* Add 10 µL of RNAse A (2 mg/mL) and mix by inverting.

5. Incubate samples in a thermomixer (or similar equipment) at 55 °C for 1 h. Then centrifuge for 5 min at maximum speed and transfer 1 mL of aqueous phase to a 2 mL microtube (in duplicate).

6. Add 500 µL of 24:1 chloroform–isoamyl alcohol and mix well the tubes with hands (by inversion), until a homogeneous emulsion is formed.

7. Centrifuge for 8–10 min at maximum speed (12,000–13,000 rpm or 9600–11300 relative centrifugal force). Following centrifugation, you should have three layers. Top: aqueous phase; middle: debris and proteins; and bottom: chloroform with cellular debris. Proceed to the next step quickly, so the phases do not remix.

8. Pipette off the aqueous phase (top) taking care not to suck up any of the middle or chloroform phases.

9. Place the aqueous phase into a new labeled 1.5 mL tube.

10. Add 30 µL of cold 7.5 M ammonium acetate.

11. Add 200 µL of cold isopropanol (−20 °C, to the combined volume of aqueous phase and added ammonium acetate).

12. Mix well by inversion.

13. Let sit in freezer for 45 min to an hour.

14. Centrifuge for 3 min at maximum speed. Orient tubes in an equal fashion to facilitate subsequent removal of supernatant without disturbing resultant DNA pellet.

15. Pipette off the liquid, being careful not to lose the pellet with your DNA. The DNA pellet at this stage is very loose and difficult to see.

16. Add 700 μL of 70% ethanol at −20 °C and mix by gentle inversion.

17. Centrifuge for 1 min at maximum speed.

18. Pipette off the liquid, being careful not to lose the pellet with your DNA.

19. Add 700 μL of 95% ethanol at −20 °C and invert to mix.

20. Centrifuge for 1 min at maximum speed.

21. Pipette off the liquid, being careful not to lose the pellet with your DNA.

22. Dry the pellet in a vacuum centrifuge or on a hot plate at 55 °C.

23. Resuspend the DNA of the samples with 100 μL of water of molecular biology grade. Store the DNA at −20 °C until required. The DNA can be stored at 4 °C if used in the next few days, or in the freezer at −20 °C for long-term storage.

24. The DNA obtained from the samples must be quantified with a spectrophotometer or similar equipment (for instance NanoDrop), and the concentration must be normalized at 100 ng/μL or multiple, to use in the following steps. The purity of the DNA solution must be higher than 1.8 measured as the ratio of absorbance at 260 and 280 nm (*see* **Note 1**).

3.2   Real-Time PCR

While PCR reactions are being set up, precautions should be taken at all stages to avoid contamination of samples and reagents. It is advisable to work in laboratories with unidirectional workflow, and at least the following rooms: DNA extraction room; master mix preparation room (in this room only reagents are handled, and no DNA enters, in order to avoid contamination of reagents); PCR room (in this room, the DNA is dispensed, and the master mix previously prepared is added); PCR room, where the thermocycler and PCR reactions are performed (*see* **Note 2**).

1. Mix well all reagents by inverting the tubes a number of times or for small volumes by flicking the tube and spin briefly (~5 s) in a centrifuge. The probe must be prepared as described in the **Note 3**.

2. Prepare the PCR master mix in 1.5 mL microcentrifuge tubes, as shown in Table 2 to maximize uniformity and minimize labor when working with multiple tubes. The final volume of each master mix depends on the number of reactions required for each real-time PCR. Master mixes can be made in excess of what is actually needed to account for pipetting variations, i.e., prepare 10% excess of master mix.

3. Mix the master mix thoroughly and dispense equal aliquots into each thin-walled PCR tube. Set up the real-time PCR in a

Table 2
Composition of PCR reactions

Mix components	Mix 1	Mix N
TaqMan® Universal PCR Master Mix (2×)	10 μL	10 × N μL
Forward primer 10 μM	1.8 μL	1.8 × N μL
Reverse primer 10 μM	1.8 μL	1.8 × N μL
Probe 10 μM	0.2 μL	0.2 × N μL
Molecular biology grade water	to 25 μL	to 25 × N μL

Table 3
PCR amplification profile of the real-time PCR system herein described

Step	Time (s)	Temperature (°C)
Hold	30	95
PCR (× 45 cycles)		
Denaturation	5	95
Annealing	30	58

96-well plate, including nontemplate (NTC) (*see* **Note 4**), negative controls, positive controls (*see* **Note 5**) and the test sample (*see* **Note 6**).

4. Add DNA template (200 ng of DNA per sample to each well) to each reaction tube as needed (be sure to include a negative control that does not receive any template). When pipetting is finished, seal the plate with optical adhesive film.

5. Centrifuge the reaction plate at low speed for 2 min and then load the plate onto a Viia7 real-time PCR System.

3.3 Program and Run Applied Biosystems ViiA 7 system

1. Operate the Applied Biosystems ViiA 7 System and SDS (Sequence Detection System) software according to the manufacturer's recommendations. Set reaction volume to 25 μL, and the amplification program as described in Table 3 (unless other parameters are established in the methodological validation). Use FAM/TAMRA as detector/quencher.

2. Briefly, launch the SDS software and open a new plate template window to denote well locations on the 96-well plate for the controls, and testing samples. Save the template window with the recorded data as an SDS run file.

3. Select the "Connect" button, and then load the 96-well plate into the instrument. Select the "Start" button to begin the actual run.

4. Run the PCR reaction to determine Ct (cycle threshold) values for each sample.

3.4 Data Collection and Analysis

After the run is completed, select baseline and place the threshold line at the exponential phase of amplification. Remove the plate from the instrument and save the results of the run. When the analysis button is selected in the SDS software, the results will be analyzed automatically if the standards and testing sample information has been recorded as noted above. The Ct values obtained should be reviewed (*see* **Notes** 7 and **8**) and interpretation of the results must be made in relation to data obtained in the internal validation performed in each laboratory for a given detection limit (*see* **Notes 9–11**), allowing to determine whether a sample contains or does not contain soy.

4 Notes

1. The quantity and purity of DNA extracted must be appropriate and sufficient to assess the presence of soy in a sample. If the amount is not enough (less than 200 ng) or the purity is not adequate, the DNA must be extracted again.

2. Precautions should be taken while PCR reactions are being set up to avoid contamination of samples and reagents. Use only dedicated pipettes for preparing master mixes and another set for adding template DNA. Use only filter tips and sterile plasticware. Wear gloves and change them frequently. Master mix preparation and addition of template should be performed in separate places with dedicated laboratory equipment. It is paramount to work in at least three separate areas. One area should be designated just for DNA extraction; other for the preparation of the master mix; and other area designated just for adding the template. This is crucial to avoid contamination of the real-time PCR reagents with aerosols containing the template. Pipette sets must be dedicated for use in these areas and must not be moved between areas.

3. TaqMan probe and probe dilution: probes are usually received as a 100 pmol/μL stock solution. This stock solution can be diluted with molecular biology grade water to achieve a working solution concentration. Probe working solutions should be aliquoted in volumes that are sufficient for one time use. These aliquots should be stored at −20 °C and protected from light. In addition, repeated freeze–thaw cycles of probe solutions should be avoided.

4. Negative reaction control (NTC) produces amplification. This may be due to amplicon contamination in the master mix reagents or, in some cases, contamination in the DNA preparation. It is often best to simply repeat the real-time PCR reaction using new reagents (except DNA prep). If the problem disappears, one can proceed. If the problem persists, a new DNA extraction is necessary.

5. The positive control is a tube of DNA extracted from a specific food matrix contaminated with soybean reference material at the detection limit of the technique. The reference material used may be Soya Bean Powder from Sigma (reference ERMBF410BK), diluted in other certified materials as for instance corn, or alternatively food matrices that do not contain soybean.

6. Test samples should be analyzed in duplicate or triplicate according to individual laboratory statistical requirements.

7. Excessive variability between duplicate or triplicate samples: If the samples display large variability between duplicates and triplicates, there may be a problem in the preparation of the master mix or a pipetting error. Check that the master mix has been prepared properly and well mixed, and that the micropipette is calibrated and is dispensing the proper amount of liquid reproducibly.

8. To consider a test as valid, it must have a correct reading of the controls (the positive control in the detection limit must be positive, and the NTC and negative controls should not have amplification).

9. The detection limit should be calculated on the basis of the validation performed on each laboratory, as it will depend on the equipment and technical personnel. Similarly, the cutoff or Ct, which separates a positive sample of a negative will vary depending on the specific detection limit of each particular laboratory.

10. To consider a curve as positive, this must have an exponential appearance, and the threshold cycle must be less than the Ct determined in the methodological validation.

11. Before applying these techniques in a particular laboratory, internal validations should be made taking into account the following points:

    Thermocyclers: As previously mentioned, each thermocycler has its own characteristics that may require an adjustment in the working parameters (temperature or reaction times, as well as temperature ramp rate).

    DNA polymerase enzyme: It is also necessary to make a validation if the polymerase employed is different from the herein described, since polymerases from different commercial dealers may vary in amplification efficiency, and consequently may affect the optimum temperature and the specificity of the technique, generating variations in observed Ct and amplification profile.

Substances included in the validation: In the phase of methodological design and development of the detection systems a great number of substances must be included for evaluating the specificity of the technique, and so validate the correct working. However, each laboratory should perform an internal validation including all the matrix used in their production chain, since these could interfere with the results, giving rise to for instance false positives.

References

1. Belloque J, García MC, Torre M, Marina ML (2002) Analysis of soyabean proteins in meat products: a review. Crit Rev Food Sci Nutr 42(5):507–532
2. Hoogenkamp HW (2005) Soy protein and formulated meat products. CABI Pub, Cambridge
3. Codex Alimentarius (2016.) http://www.fao.org/fao-who-codexalimentarius/codex-home/es/. Accessed Jun 2016.
4. Espiñeira M, Herrero B, Vieites JM, Santaclara FJ (2010) Validation of end-point and real-time PCR methods for the rapid detection of soy allergen in processed products. Food Addit Contam Part A 27(4):426–432
5. Zhang M, Gao X, Yu Y, Ao J, Qin J, Yao Y, Li Q (2007) Detection of roundup ready soy in highly processed products by triplex nested PCR. Food Control 18(10):1277–1281

Chapter 13

RAPD/SCAR Approaches for Identification of Adulterant Breeds' Milk in Dairy Products

Joana T. Cunha and Lucília Domingues

Abstract

Food safety and quality are nowadays a major consumers' concern. In the dairy industry the fraudulent addition of cheaper/lower-quality milks from nonlegitimate species/breeds compromises the quality and value of the final product. Despite the already existing approaches for identification of the species origin of milk, there is little information regarding differentiation at an intra-species level. In this protocol we describe a low-cost, sensitive, fast, and reliable analytical technique—Random Amplified Polymorphic DNA/Sequence Characterized Amplified Region (RAPD/SCAR)—capable of an efficient detection of adulterant breeds in milk mixtures used for fraudulent manufacturing of dairy products and suitable for the detection of milk adulteration in processed dairy foods.

Key words DNA extraction, RAPD, SCAR, Breed identification, Milk adulteration, Dairy authentication, Food quality control

1 Introduction

The increase of consumer's awareness about food safety and quality has resulted in a growing attention toward the improvement and establishment of food authentication processes, mainly concerning high-quality products that, having an elevated economic value, are more prone to adulterations, with severe effects in the agro-food industry and market. In the dairy industry, one of the major fraudulent activities consists in the addition of cheaper and/or low quality milk from species/breeds other than the ones that should be used for a legitimately manufacture of a specific product. These adulterations will almost certainly cause a negative impact in the quality of the final product, leading to serious problems at both social and economic levels, creating an urgent necessity for robust and reliable analytical techniques that efficiently allow the identification of milk origin in processed dairy products. Since the discovery that the somatic cells present in milk may be used as source of DNA [1], several PCR-based techniques have been developed to

Lucília Domingues (ed.), *PCR: Methods and Protocols*, Methods in Molecular Biology, vol. 1620,
DOI 10.1007/978-1-4939-7060-5_13, © Springer Science+Business Media LLC 2017

perform species identification of milk in dairy products, such as multiplex [2–4], Restriction Fragment Length Polymorphism (RFLP) [5], and Real-Time PCR [2, 6, 7]. However, the differentiation of milk source at an intraspecies level is still a major challenge [8–10]. Furthermore, the lack of genetic information at a breed level poses as a major obstacle, creating the necessity for more specific strategies.

Taking these into account, in this chapter we describe a Random Amplified Polymorphic DNA/Sequence Characterized Amplified Regions (RAPD/SCAR) protocol suitable for identification of adulterant breeds' milk in dairy products. Figure 1 schematically shows the steps of this procedure. The RAPD/SCAR approach is ideal for studies where none or little information about the target genome is available, as the RAPD analysis does not require any prior DNA knowledge of the organism under study. This technique consists in using small arbitrary primers that will hybridize randomly to the target DNA, ideally resulting in a breed-specific band pattern. For the specific objective of identifying fraudulent addition of milk in dairy products, the patterns obtained from putative adulterant breeds should be compared with the patterns obtained from the legitimate breed, for identification of distinguishing fragments. Despite the skepticism over reproducibility in RAPD analysis, several studies suggest that an optimized RAPD protocol, when performed in strictly consistent conditions and focused in fragments with satisfactory repeatability, may achieve reproducible results [11–13] and be a valuable tool for identification

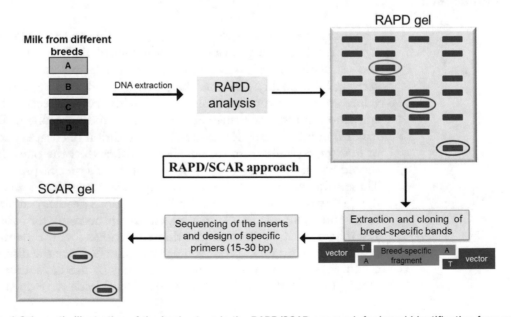

Fig. 1 Schematic illustration of the basic steps in the RAPD/SCAR approach for breed identification from milk samples

of sheep breed in a first stage of milk authentication in dairy industry [10]. The SCAR technique is developed based on results obtained from the RAPD analysis, allowing the design of longer and more specific primers (using the breed-specific fragments identified by RAPD pattern analysis as template), suitable for breed detection in products presenting DNA degradation resulting from the milk processing in the dairy industry. In this sense, when combined, the RAPD and SCAR analyses pose as a potential tool for the detection of milk origin in processed dairy products.

2 Materials

2.1 Equipment

1. Microtubes (0.2 mL PCR tubes and 1.5 mL microcentrifuge tubes).

2. Pipettes and pipette tips.

3. Microcentrifuge.

4. Incubator with orbital shaking.

5. Fume hood.

6. Microvolume UV–Vis spectrophotometer for nucleic acid quantification.

7. Thermal cycler.

8. Microwave.

9. Agarose gel electrophoresis apparatus.

10. UV transilluminator with a gel documentation unit.

11. Flow chamber.

12. Stationary incubator.

2.2 Supplies and Reagents

1. Milk samples, stored at −20 °C to prevent contamination and degradation.

2. Clearing Solution: 0.25 M EDTA, pH 8.0, 0.5% Triton X-100.

3. Extraction Buffer: 10 mM Tris–HCl, pH 7.5, 150 mM NaCl, 2 mM EDTA, pH 8.0, and 1% SDS.

4. 5 M guanidine hydrochloride.

5. 20 mg/mL Proteinase K.

6. Phenol–chloroform–isoamyl alcohol (25:24:1).

7. Absolute ethanol.

8. Sterile ultrapure water.

9. Oligonucleotide primers: Ordered from a local supplier at a 20 nmol synthesis scale, and without HPLC or gel purification. Aliquots of 20 μM working solution were prepared from 100 μM stock solution and stored at −20 °C to prevent contamination of stock and repeated freeze–thaw cycles.

10. Taq DNA polymerase: In this protocol we used NZYTaq DNA polymerase, supplied with a 10× Reaction buffer and a 50 mM MgCl$_2$ solution (NZYtech).

11. dNTPs, 10 mM each.

12. Dimethyl sulfoxide (DMSO).

13. 1× Tris–Borate–EDTA (TBE) buffer: 89 mM Tris (pH 7.6), 89 mM boric acid, 2 mM EDTA.

14. 1.5% (w/v) agarose gels (in 1x TBE buffer) with 0.05% (v/v) GreenSafe Premium (NZYTech), or equivalent.

15. Molecular weight marker, NZYDNA Ladder VII or equivalent.

16. 6× DNA loading buffer: 30% (w/v) glycerol, 0.25% (w/v) bromophenol blue.

17. QIAquick Gel Extraction Kit (Qiagen) or equivalent.

18. Commercial systems for cloning of PCR products: In this protocol we used pGEM-T Easy vector, supplied with a 2X Rapid Ligation Buffer and T4 DNA Ligase (Promega).

19. *Escherichia coli* chemically competent cloning cells: In this protocol we use NZY5α Competent Cells (NZYTech).

20. SOC medium: 2% tryptone, 0.5% yeast extract, 10 mM NaCl, 2.5 mM KCl, 10 mM MgCl$_2$, 10 mM MgSO$_4$, and 20 mM glucose.

21. LB agar plates: 1% tryptone, 0.5% yeast extract, 1% NaCl, 2% agar, with appropriate antibiotics.

22. GenElute Plasmid Miniprep Kit (Sigma-Aldrich), or equivalent.

3   Methods

3.1 DNA Extraction from Milk Samples

1. Harvest somatic cells by mixing 0.5 mL of Clearing Solution with 1 mL of milk sample and centrifuging at 16,100 g for 5 min. Remove the resulting cream pad and supernatant (*see* **Note 1**).

2. Lyse the somatic cells by ressuspending the pellet in 860 μL of Extraction Buffer, 100 μL of 5 M guanidine hydrochloride and 20 μL of 20 mg/mL Proteinase K. Incubate for 3 h or o/n at 55 °C, 80 rpm.

3. Let the cell lysate cool at room temperature. In a fume hood add 500 μL of phenol–chloroform–isoamyl alcohol (25:24:1) and centrifuge at 16,100 g for 15 min. Collect the clear aqueous upper phase containing DNA (*see* **Note 2**).

4. Precipitate DNA by adding 800 μL of ice-cold absolute ethanol and centrifuge at 4 °C and 16,100 rpm for 30 min. Discard the supernatant and let the pellet to dry in a fume hood (*see* **Note 3**).

5. Elute the DNA in 30 μL of sterile ultrapure water. Determine spectrophotometrically the DNA concentration and purity (*see* **Note 4**).

3.2 RAPD Analysis

1. Select appropriate RAPD primers from the literature (*see* **Note 5**).

2. Select DNA from one individual of each breed to perform a preliminary analysis with all the RAPD primers (*see* **Note 6**).

3. For the RAPD PCR, prepare the following reaction mixture (*see* **Note 7**).

   - Template DNA: 200 ng.
   - 50 mM MgCl$_2$: 4 μL.
   - 10 mM dNTPS: 2.5 μL.
   - 20 μM RAPD primer: 2.5 μL.
   - 100% DMSO: 2.5 μL.
   - 5 U/μL Taq polymerase: 0.5 μL.
   - 10x Reaction buffer: 5 μL.
   - Ultrapure H$_2$O: up to 50 μL (depending on template DNA concentration).

4. Perform the amplification in a thermal cycler programmed as follows (*see* **Note 7**):

Number of cycles	PCR step	Temperature (°C)	Time (min)
1	Initial denaturation	95	2
45	Denaturation	95	1
	Annealing	34	1
	Extension	72	2
1	Final extension	72	5

5. Load the amplification products on 1.5% agarose gel in TBE and run in the same buffer for 2.5 h at 100 V (*see* **Note 8**).

6. Compare the resulting RAPD pattern, searching for fragments present in the adulterants breed and absent in the original breed pattern. Figure 2 shows an example of the RAPD amplification patterns obtained when testing different breeds.

7. Select RAPD primer(s) that resulted in breed-specific bands to test in a larger number of individuals of each breed (repeat **steps 3–5**).

8. Analyse the resulting patterns to check for intrabreed consistency, i.e., breed-specific bands must be present in all of the individuals tested. Figure 3 shows an example of the RAPD amplification patterns obtained when testing different individuals from each breed. Repeat RAPD amplifications to check for intralaboratory technique consistency.

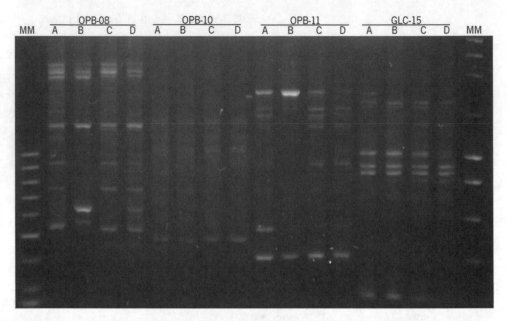

Fig. 2 RAPD amplification patterns obtained, with primers OPB-08 (5′ GTCCACACGG 3′), OPB-10 (5′ CTGCTGGGAC 3′), OPB-11 (5′ GTAGACCCGT 3′), and GLC-15 (5′ GACGGATCAG 3′), from one individual of each of the four different sheep breeds (A, B, C, and D). MM, molecular marker

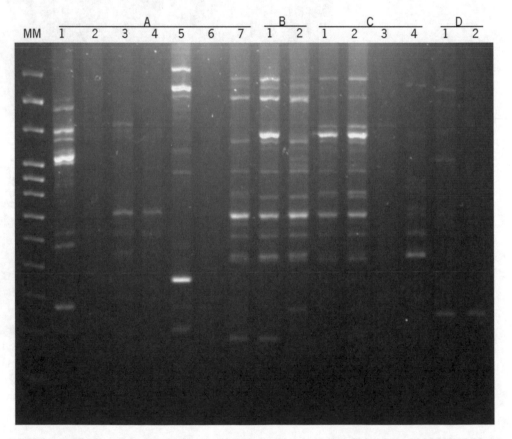

Fig. 3 RAPD amplification patterns obtained, with two combined primers GLA-19 (5′ CAAACGTCCGG 3′) and OPB-12 (5′ CCTTGACGCA 3′), from several individuals of each breed: seven of A, two from B, four from C, and two from D. MM, molecular marker

3.3 SCAR Analysis

After identification of an adulterant-specific fragment in the RAPD analysis (*see* **Note 9**), SCAR primers may be designed to specifically amplify that sequence region.

1. Excise the adulterant-specific fragment from the agarose gel and purify it using a QIAquick Gel Extraction Kit, or equivalent, and clone the obtained fragment into the pGEM-T Easy vector, or similar, accordingly to the supplier's instructions (*see* **Note 10**).

2. Transform *E. coli* chemically competent cells, such as NZY5α Competent Cells, following manufacturer's instructions (*see* **Note 11**). Plate transformed cells in LB agar plates containing appropriated antibiotics (*see* **Note 11**) and incubate overnight at 37 °C. Confirm the correct insertion of the desired fragment by colony PCR (using the RAPD primer that previously generated this fragment and the reaction described in Subheading 3.2, **steps 3** and **4**) or restriction analysis.

3. Extract the plasmid DNA from 5 clones containing the expected insert using GenElute Plasmid Miniprep Kit, or equivalent, according to supplier's instructions. Send the resulting purified fragments for sequencing in a local services provider (such as GATC Biotech).

4. Confirm that the different clones contain equal sequences of the breed-specific fragment. Search databases for sequence homology using Blast program (http://blast.ncbi.nlm.nih.gov/Blast.cgi) (*see* **Note 12**).

5. Based in the sequence of the adulterant-specific fragment, design SCAR primers with stringent conditions of size (15–30 bp), GC (guanine–cytosine) percentage (40–60%), and melting temperature (52–65 °C, and less than 5 °C of difference within primer pair). Check for potential primer dimer and hairpin formation using computational tools such as the online OligoAnalyzer tool (Integrated DNA Technologies).

6. Optimize the amplification reaction in terms of reagents' concentration (e.g., DNA, primers, $MgCl_2$) and of reaction temperatures and duration, until a clear, intense, and distinguishable band is observed when amplifying DNA from the adulterant breed in focus. Presented below is an example of the reaction mixture and thermalcycling program resulting of the optimization performed for the SCAR primers designed for a sheep breed [10].

- Template DNA: 100 ng.
- 50 mM MgCl2: 3 μL.
- 10 mM dNTPS: 2.5 μL.
- 2 μM SCAR primer forward: 2.5 μL.
- 2 μM SCAR primer forward: 2.5 μL.
- 5 U/μL Taq polymerase: 0.5 μL.
- 10x Reaction buffer: 5 μL.
- Ultrapure H2O: up to 50 μL (depending on template DNA concentration).

Number of cycles	PCR step	Temperature (°C)	Time
1	Initial denaturation	95	2 min
45	Denaturation	95	1 min
	Annealing	63	1 min
	Extension	72	20 s
1	Final extension	72	5 min

7. Test milk from several individuals of different breeds with these SCAR primers and the optimized reaction conditions. Resolve the PCR products by electrophoresis on 1.5% agarose gel in TBE for 1 h at 100 V. Amplification should only be observed from DNA of the adulterant breed from which the SCAR markers originated.

4   Notes

1. The quantity of somatic cells in milk may vary depending on the species or breed of the source organism or even on the period of milk collection. Taking this into account a volume larger than 1 mL may be necessary to attain satisfactory concentrations of DNA (*see* **Note 4**).

2. The phenol–chloroform–isoamyl alcohol protocol for DNA extraction from lysed cells results in the formation of two separate phases: a lower organic phase, containing denatured proteins, and the upper phase, containing nucleic acids. Caution should be used when collecting the aqueous upper phase to avoid cross contamination between phases that may negatively affect the subsequent PCRs.

3. Absolute ethanol should be kept at −20 °C before use, as the low temperature decreases DNA solubility and allows a more rapid and efficient precipitation. It should be noted that, depending on its concentration, the DNA pellet may not be

visible or appear only as thin smear, so the supernatant removal should be carefully performed to avoid accidental discard of DNA.

4. The extraction protocol will inevitably result in DNA solutions containing contaminants that may inhibit PCR, so we suggest selection of DNA samples with concentrations superior to 30 ng/μL, as higher concentrations require a minor addition of DNA solution to the PCR mixture, and consequently of a smaller presence of contaminants. For the same reason we suggest selecting samples with purity ratios of $A_{260/280}$ superior to 1.6 and $A_{260/230}$ superior to 1.0, as smaller ratios indicate high contamination with protein or organic compounds/chaotropic salts, respectively.

5. RAPD primers may be obtained from commercially available kits. Alternatively, RAPD primers may be selected from studies where they have been successfully used for breed identification/differentiation or for genetic diversity determination.

6. Perform the analysis using the RAPD primers individually or, in case of low polymorphism observed, perform amplifications with a combination of 2–3 RAPD primers.

7. The reaction mixture and thermal cycler programme were optimized for attainment of a clear RAPD banding pattern. Different reaction mixture volume and concentrations of template DNA, primer, $MgCl_2$, and DMSO, as well as number of PCR cycles and annealing temperature, were tested until achievement of RAPD patterns with a satisfactory number of distinct bands.

8. The large number of bands in the RAPD pattern and their small size create the necessity for an higher resolution of separation, so gels should be prepared with higher concentrations of agarose (such as 1.5% (w/v)) and in TBE buffer (borate resolves <2 kb fragments more efficiently than acetate). Additionally, the TBE buffer, as a highly conductive medium, is ideal to prevent overheating in long electrophoretic runs.

9. Selecting smaller bands from the RAPD patterns for SCAR analysis is recommended, as a small predicted amplicon may offer a further advantage when analyzing processed dairy products with partially degraded DNA.

10. The pGEM-T Easy vector uses the TA cloning strategy, so the fragment used for insertion needs to present A-overhangs, i.e., be amplified using a Taq DNA polymerase. If other type of DNA polymerase is used for the RAPD analysis, resulting in blunt-end fragments, the cloning strategy must be revised (e.g., using commercial kits designed to clone blunt PCR fragments) or poly(A) tails may be added to the fragment, by adding 1 U

of Taq DNA polymerase, 1x Reaction buffer, and 0.2 mM of dATPs, and performing a 10-min extension step at 72 °C.

11. The ligation reaction and transformation using the pGEM-T Easy vector and NZY5α Competent Cells is highly efficient, so it is possible to use a smaller amount of competent cells (up to one fifth) for each transformantion reaction and attain a satisfactory number of colonies. Additionally the pGEM-T Easy vector allows for blue/white screening of recombinants on LB plates containing ampicillin, IPTG, and X-gal.

12. If the fragment's sequence presents homology with an annotated gene of the organism in study it is possible that its identification in the RAPD patterns was due to a point mutation, meaning that this fragment is not breed-specific and therefore should not be used as base for the design of SCAR primers. Additionally, searching for homology with lactic acid bacteria may eliminate the possibility of contamination of the milk samples with these bacteria (if lactic acid bacteria are present in the milk samples, their DNA would also be extracted and contaminate the DNA from the somatic cells, which may result in artifacts in the RAPD analysis).

Acknowledgments

This study was supported by the Portuguese Foundation for Science and Technology (FCT) under the scope of the strategic funding of UID/BIO/04469/2013 unit and COMPETE 2020 (POCI-01-0145-FEDER-006684), BioTecNorte operation (NORTE-01-0145-FEDER-000004) funded by European Regional Development Fund under the scope of Norte2020—Programa Operacional Regional do Norte, and project "Valor Queijo" (CENTRO-07-0202-FEDER-030372) funded by FCT/ MCTES (PIDDAC) and cofunded by "Fundo Europeu de Desenvolvimento Regional" (FEDER) through "COMPETE—Programa Operacional Factores de Competitividade" (POFC).

References

1. Lipkin E, Shalom A, Khatib H et al (1993) Milk as a source of deoxyribonucleic acid and as a substrate for the polymerase chain reaction. J Dairy Sci 76:2025–2032

2. Agrimonti C, Pirondini A, Marmiroli M et al (2015) A quadruplex PCR (qxPCR) assay for adulteration in dairy products. Food Chem 187:58–64

3. Bottero MT, Civera T, Nucera D et al (2003) A multiplex polymerase chain reaction for the identification of cows', goats' and sheep's milk in dairy products. Int Dairy J 13:277–282

4. Gonçalves J, Pereira F, Amorim A et al (2012) New method for the simultaneous identification of cow, sheep, goat, and water buffalo in dairy products by analysis of short species-specific mitochondrial DNA targets. J Agric Food Chem 60:10480–10485

5. Abdel-Rahman SM, Ahmed MMM (2007) Rapid and sensitive identification of buffalo's,

cattle's and sheep's milk using species-specific PCR and PCR-RFLP techniques. Food Control 18:1246–1249

6. Ganopoulos I, Sakaridis I, Argiriou A et al (2013) A novel closed-tube method based on high resolution melting (HRM) analysis for authenticity testing and quantitative detection in Greek PDO feta cheese. Food Chem 141: 835–840

7. López-Calleja I, González I, Fajardo V et al (2007) Quantitative detection of goats' milk in sheep's milk by realtime PCR. Food Control 18:1466–1473

8. Fontanesi L, Beretti F, Dall'Olio S et al (2011) A melanocortin 1 receptor (MC1R) gene polymorphism is useful for authentication of Massese sheep dairy products. J Dairy Res 78:122–128

9. Sardina MT, Tortorici L, Mastrangelo S et al (2015) Application of microsatellite markers as potential tools for traceability of Girgentana goat breed dairy products. Food Res Int 74: 115–122

10. Cunha JT, Ribeiro TIB, Rocha JB et al (2016) RAPD and SCAR markers as potential tools for detection of milk origin in dairy products: adulterant sheep breeds in Serra da Estrela cheese production. Food Chem 211:631–636

11. Blixt Y, Knutsson R, Borch E et al (2003) Interlaboratory random amplified polymorphic DNA typing of *Yersinia enterocolitica* and *Y. enterocolitica*-like bacteria. Int J Food Microbiol 83:15–26

12. Güçlü F, Aldem OS, Güler L (2004) Differential identification of cattle *Sarcocystis spp.* by random amplified polymorphic DNA polymerase chain reaction (RAPD-PCR). Rev Med Vet 155:440–444

13. Perry AL, Worthington T, Hilton AC (2003) Analysis of clinical isolates of *Propionibacterium acnes* by optimised RAPD. FEMS Microbiol Lett 228:51–55

Chapter 14

Genetic Diversity Analysis of Medicinally Important Horticultural Crop *Aegle marmelos* by ISSR Markers

Farina Mujeeb, Preeti Bajpai, Neelam Pathak, and Smita Rastogi Verma

Abstract

Inter simple sequence repeat (ISSR) markers help in identifying and determining the extent of genetic diversity in cultivars. Here, we describe their application in determining the genetic diversity of bael (*Aegle marmelos* Corr.). Universal ISSR primers are selected and their marker characteristics such as polymorphism information content, effective multiplex ratio and marker index have been evaluated. ISSR-PCR is then performed using universal ISSR primers to generate polymorphic bands. This information is used to determine the degree of genetic similarity among the bael varieties/accessions by cluster analysis using unweighted pair-group method with arithmetic averages (UPGMA). This technology is valuable for biodiversity conservation and for making an efficient choice of parents in breeding programs.

Key words Bael (*Aegle marmelos*), DNA fingerprinting, Genetic diversity, ISSR, UPGMA

1 Introduction

Knowledge of genetic variation and genetic relationship among genotypes is an important consideration for classification, utilization of germplasm resources and breeding. Morphological traits for the classification of plants, when used for determining diversity and relationships among plants species, are not sufficient because of environmental influences on them. Thus, the usefulness of molecular markers has been investigated as a means of characterizing and discriminating different species more precisely [1]. The introduction of molecular biology techniques, such as DNA-based markers, allows for direct comparison of different genetic materials independent of environmental influences. The viability and purity of accessions can be analyzed through the utilization of fingerprints based on molecular markers. This process can increase both quantity and quality of plant. Studies with molecular markers have made significant contributions to our understanding of genetic diversity. When compared with other types of markers, they present a greater number of polymorphic loci, which allows

Lucília Domingues (ed.), *PCR: Methods and Protocols*, Methods in Molecular Biology, vol. 1620,
DOI 10.1007/978-1-4939-7060-5_14, © Springer Science+Business Media LLC 2017

distinguishing between accessions that may have similar morphological and agronomical traits [2]. DNA markers are thus being widely used in studying polymorphism between species or in populations. The degree of similarity between the banding patterns provides information about genetic similarity and relationships between the samples studied. The application largely depends on the type of markers employed, distribution of markers in the genome, type of loci they amplify, level of polymorphism and reproducibility of products [3, 4]. The introduction of DNA markers based on the polymerase chain reaction (PCR) technology has led to the development of several novel genetic assays that can be used in plant genetic analysis such as cultivar identification and gene mapping. Among various molecular tools for genetic analysis, inter simple sequence repeat (ISSR) markers, which involve PCR amplifications of DNA using a primer composed of a microsatellite sequence anchored at the 3' or 5' end by 2–4 arbitrary nucleotides, are a powerful tool for investigating genetic variation within species [5, 6].

The ISSR markers have been reported to produce more complex marker patterns than the random amplification of polymorphic DNA (RAPD) approach, which is advantageous when differentiating closely related cultivars [7]. ISSR markers produce a much greater number of total polymorphic and discriminating fragments [8], and require lower relative costs compared with RAPDs. In addition, ISSR markers are more reproducible than RAPD markers, because ISSR primers designed to anneal to a microsatellite sequence, are longer than RAPD primers, allowing higher annealing temperatures to be used [9]. As compared to amplified fragment length polymorphism (AFLP), ISSR detection is an easier and cheaper technique [10]. The ISSR technique is simpler to use than the SSRs [11, 12] and less restrictive than RFLPs, and may offer considerable variation among species [13]. Owing to its advantages, ISSR marker profiling is increasingly used for various applications, for example, DNA fingerprinting, phylogenetic and population genetic diversity analysis in many plant species [14–19]. ISSR markers have been used successfully in a number of horticultural crops including potato, chestnut, strawberry, cashew, mulberry, and *Clematis* species [20–25]. Till date, no report is available on application of ISSR markers in genetic diversity analysis of *Aegle marmelos* (bael).

2   Materials

Obtain several bael genotypes/accessions/varieties and authenticate the taxonomy of the plant. Prepare all reagents in ultrapure sterile Milli-Q water using analytical grade or molecular biology grade chemicals.

2.1 Plant DNA Isolation

1. 1 M Tris (hydroxymethyl) aminomethane, pH 8.0 (Tris–HCl). To prepare 1 M stock solution of Tris–HCl, pH 8.0, measure about 60 mL of water with the help of a graduated cylinder and transfer it to a glass beaker. Weigh 15.76 g of Tris and transfer to the beaker containing water. Adjust the pH with 1 M HCl (*see* **Note 1**). Make up the volume to 100 mL with water. Sterilize by autoclaving at 15 psi for 15 min. Store at 4 °C. Take appropriate volume of 1 M stock solution to get a final concentration of 100 mM in the extraction buffer (*see* **Note 2**).

2. 0.5 M Ethylene diamine tetraacetic acid, disodium salt (Na$_2$EDTA), pH 8.0. To prepare a 100 mL solution of 0.5 M stock solution of Na$_2$EDTA, pH 8.0, weigh 18.61 g of Na$_2$EDTA and transfer it to a glass beaker. Initially add 50 mL water. Adjust the pH at 8.0 with NaOH. Make up the volume to 100 mL with water (*see* **Note 3**). Sterilize by autoclaving at 15 psi for 15 min and store at 4 °C. Take appropriate volume of this 0.5 M stock solution to get a final concentration of 20 mM in the extraction buffer (*see* **Note 2**).

3. 5.0 M Sodium chloride (NaCl). Add 29.23 g of NaCl to a beaker containing 50 mL of water. Stir well and make up the volume to 100 mL with water. Sterilize by autoclaving at 15 psi for 15 min. Take appropriate volume of this 5 M stock solution to get final concentration of 2 M in the extraction buffer (*see* **Note 2**).

4. Cetyl trimethyl ammonium bromide (CTAB): Add 2% (w/v) CTAB in extraction buffer.

5. 1.5% (w/v) polyvinylpyrrolidone (PVP) (*see* **Note 4**).

6. 2.5% (v/v) β-mercaptoethanol (β-ME) (*see* **Note 4**).

7. 10 mg/mL ribonuclease A (RNase A). Weigh 10 mg of RNase A and transfer to a 1.5 mL microfuge tube. Add 0.5 mL of DEPC treated nuclease and protease free water, mix well to dissolve and make up the volume to 1 mL. Store at 4 °C.

8. Chloroform–isoamyl alcohol (24:1). Mix 48 mL of chloroform with 2 mL of isoamyl alcohol.

9. Isopropanol. Store at −20 °C.

10. 70% (v/v) ethanol. Mix 70 mL of absolute ethanol with 30 mL of sterile distilled water. Store at −20 °C.

11. TE buffer (pH 8.0). Buffer containing 10 mM Tris–HCl (pH 8.0), 1 mM Na$_2$EDTA (pH 8.0). For preparing 10 mL TE buffer, add 100 μL of 1 M Tris–HCl (pH 8.0) and 20 μL of 0.5 M Na$_2$EDTA, and make the volume up to 10 mL with sterile distilled water. Store at 4 °C.

2.2 Agarose Gel Electrophoresis, Quantification, and Visualization of DNA

1. Midi subsystem of submarine gel electrophoresis unit with power pack.

2. 50× Tris–Acetate–EDTA Buffer (TAE), pH 8.0. 50× electro- phoresis and gel buffer. For preparing 100 mL of 50× TAE buffer, add 24.2 g Tris free base (M.W. 121.14), 10 mL of 0.5 M stock solution of Na_2EDTA (pH 8.0), 5.71 mL glacial acetic acid (M.W 60) and remaining volume of sterile distilled water. For preparing 1× working solution (1× TAE), dilute 20 mL of 50× TAE to 1000 mL with distilled water. The 1× TAE solution is 40 mM Tris, 20 mM Acetate and 1 mM Na_2EDTA (*see* **Note 5**).

3. DNA intercalating dye: 10 mg/mL ethidium bromide (EtBr). Dissolve 10 mg EtBr in 1 mL of sterile distilled water (*see* **Note 6**). Store at 4 °C.

4. Agarose gel: 0.8% (w/v) and 1.5% (w/v) agarose gel for visu- alization of genomic DNA and amplified products, respec- tively. Prepare agarose gels by melting 0.4 and 0.75 g, respectively, in 50 mL of 1× TAE buffer (pH 8.0) by heating (>90 °C) in a microwave oven until completely melted. Cool molten agarose (tolerable to hand) and add the stock of EtBr at the concentration of 0.5 μg/mL before casting the gel (*see* **Notes 6** and **7**). Pour the solution on casting tray in the elec- trophoresis unit with comb placed in order and allow it to solidify at room temperature [26].

   After solidification, immerse the gel under 1× TAE buffer in the electrophoresis tank and gently remove the comb, leaving wells where DNA samples are loaded.

5. 6× Tracking or sample loading dye: 0.25% bromophenol blue (BPB), 30% glycerol. Prepare 100 mL of tracking dye by add- ing 0.25 g BPB and 30 mL glycerol in rest volume of sterile distilled water. Store at 4 °C.

6. DNA size marker: λDNA *Eco*R I + *Hin*d III double digest.

7. DNA ladder: 100 bp.

8. Gel documentation system for visualization of EtBr-stained gel.

9. NanoDrop spectrophotometer (ND2000).

2.3 PCR Reaction Mixture

1. Thermal cycler.

2. 10× PCR buffer.

3. 50 mM magnesium chloride ($MgCl_2$).

4. 10 mM deoxyribonucleoside triphosphate (dNTP) mix con- taining four types of dNTPs.

5. 1 U *Taq* DNA polymerase.

3 Methods

3.1 Plant DNA Extraction

1. Prepare 10 mL of extraction buffer by mixing appropriate volumes of respective stock solutions to get a final composition of 100 mM Tris–HCl (pH 8.0), 20 mM Na_2EDTA (pH 8.0), and 2 M NaCl. Weigh and add 2% (w/v) CTAB and make up the final volume to 10 mL with sterile distilled water (*see* **Note 8**).

2. Preheat the extraction buffer in a water bath at 65 °C for about 15 min (*see* **Note 9**).

3. Submerge 1 g of plant tissue in 5 mL of absolute alcohol for 5 min and allow alcohol to evaporate.

4. Grind the tissue in the presence of pre-warmed 10 mL extraction buffer, 1.5% (w/v) PVP and 2.5% (v/v) β-ME by using a prechilled mortar and pestle (−40 °C/−80 °C).

5. Add 6 μL of 10 mg/mL RNase to remove RNA contamination.

6. Transfer the homogenate into 2 mL centrifuge tubes and incubate in a water bath at 65 °C for 1 h.

7. Centrifuge the tubes at 10,000 rpm for 10 min at 4 °C and collect the supernatant in a 1.5 mL microfuge tube.

8. Extract the supernatant by mixing with equal volume of chloroform–isoamyl alcohol (24:1) for 15 min and centrifugation at 10,000 rpm for 10 min at 4 °C.

9. Collect the supernatant in a 1.5 mL microfuge tube and re-extract by repeating **step 8**.

10. Take the supernatant in a 1.5 ml microfuge tube and add twice the volume of chilled isopropanol to precipitate the DNA at −20 °C for 30 min (*see* **Note 10**).

11. Centrifuge the tubes at 10,000 rpm for 10 min at 4 °C and collect the pellet.

12. Wash the pellet 1–2 times with 70% (v/v) ethanol (*see* **Note 11**).

13. Air-dry the pellet at room temperature.

14. Dissolve the DNA in 100 μL of TE buffer and store at −20 °C for further use.

3.2 Quantification and Visualization of DNA

1. Quantify and determine the purity of DNA by measuring optical density (O.D.) at A_{260} and A_{280} with a NanoDrop Spectrophotometer (*see* **Notes 12** and **13**).

2. Dilute DNA to a concentration of 50 ng/μL and store at −20 °C for further use.

3. Prepare DNA samples by mixing 6X tracking dye to get a final concentration of 1× (*see* **Note 14**).

4. Load 10 μL samples on a 0.8% (w/v) agarose gel containing 0.5 μg/mL EtBr, electrophorese in 1× TAE buffer at 80 V [26] (*see* **Notes 6** and **15**).

5. Visualize and photograph under a gel documentation system (*see* **Notes 6** and **15**).

3.3 Optimization of the PCR Amplification Reaction

1. Select and purchase universal ISSR primers for PCR amplification of genomic DNA from several varieties/accessions of *Aegle marmelos* (Table 1) (*see* **Note 16**).

2. Perform PCR amplifications by using the following PCR reaction mixture: 25 μL reaction mixture containing 2.0 μL of 50 ng template DNA, 2.5 μL of 10× PCR buffer, 1.0 μL of 50 mM $MgCl_2$, 2.0 μL of 10 mM dNTPs mix, 2.0 μL of 10 picomol primers, 0.5 μL of 1 U *Taq* DNA polymerase, and 15.0 μL of sterile Milli-Q water (*see* **Note 17**).

3. Mix all the reagents by giving a short spin and amplify in a DNA thermal cycler.

Table 1
ISSR primers used in the investigation of bael varieties/accessions

S. No.	ISSR primers	Sequence (5′ → 3′)
1.	UBC 807	AGA GAG AGA GAG AGA GT
2.	UBC 810	GAG AGA GAG AGA GAG AT
3.	UBC 811	GAG AGA GAG AGA GAG AC
4.	UBC 815	CTC TCT CTC TCT CTC TG
5.	UBC 824	TCT CTC TCT CTC TCT CG
6.	UBC 825	ACA CAC ACA CAC ACA CT
7.	UBC 834	AGA GAG AGA GAG AGA GYT
8.	UBC 836	AGA GAG AGA GAG AGA GYA
9.	UBC 840	GAG AGA GAG AGA GAG AYT
10.	UBC 841	GAG AGA GAG AGA GAG AYC
11.	UBC 842	GAG AGA GAG AGA GAG AYG
12.	UBC 859	TGT GTG TGT GTG TGT GRC
13.	UBC 888	BDB CAC ACA CAC ACA CA
14.	UBC 889	DBD ACA CAC ACA CAC AC
15.	UBC 890	VHV GTG TGT GTG TGT GT
16.	UBC 891	HVH TGT GTG TGT GTG TG

where V = A, C, G; B = G, C, T; H = A, C, T; D = A, G, T; Y = pyrimidine (C, T); R = purine (A, G)

4. Thermal cycling conditions are as follows: initial denaturation step for 2 min at 94 °C, followed by 40 cycles each of 1 min at 94 °C (denaturation), 1 min at 55 °C (annealing for ISSR primers), 2 min at 72 °C (extension) followed by one final extension of 7 min at 72 °C and at last the hold temperature of 4 °C (*see* **Note 17**).

5. Prepare samples by mixing 6× tracking dye to a final concentration of 1×. Load the samples in 1.5% agarose gel containing 0.5 µg/mL EtBr. In two separate wells, also load 2 µL of 0.5 µg/µL solution (i.e., 1 µg) of DNA size marker and DNA ladder as standards (*see* **Note 18**). Electrophorese in 1× TAE at 80 V (*see* **Notes 6** and **15**).

6. Visualize and photograph the gels under a gel documentation system (*see* **Notes 6** and **15**).

7. Determine the size of fragments generated by comparing them with the DNA marker and the DNA ladder.

8. For each experiment, test the reproducibility of the amplification products twice using similar reaction conditions at different times (Fig. 1).

3.4 Analysis of Marker Properties

In order to assess the ability of the most informative primers to differentiate between the bael varieties/accessions, calculate various parameters for determining the marker efficiency and genetic characteristics (*see* **Note 19**).

1. Calculate the polymorphism information content (PIC) (Table 2).

    PIC $= 1 - p2 - q2$ [27], where p is the band frequency and q is no band frequency. Obtain the frequency of an allele by dividing the number of isolates where the band is found by the total number of isolates.

2. Calculate the effective multiplex ratio (EMR or E) (Table 2).

    E is the product of the total number of fragments per primer (n) and the fraction of polymorphic fragments (β) [20, 28]. Thus,

    $E = n\beta$, where n = total number of bands and β = total number of polymorphic bands

3. Calculate the marker index (MI) (Table 2).

    MI is the product of PIC and EMR [29], given as: MI = PIC × EMR

3.5 ISSR Data Analysis and Scoring

1. Transform unequivocally scorable and consistently reproducible amplified DNA fragments into binary data (1 = presence, 0 = absence).

2. Analyze the results on the principle that a band is "polymorphic" if it is present in some individuals and absent in others,

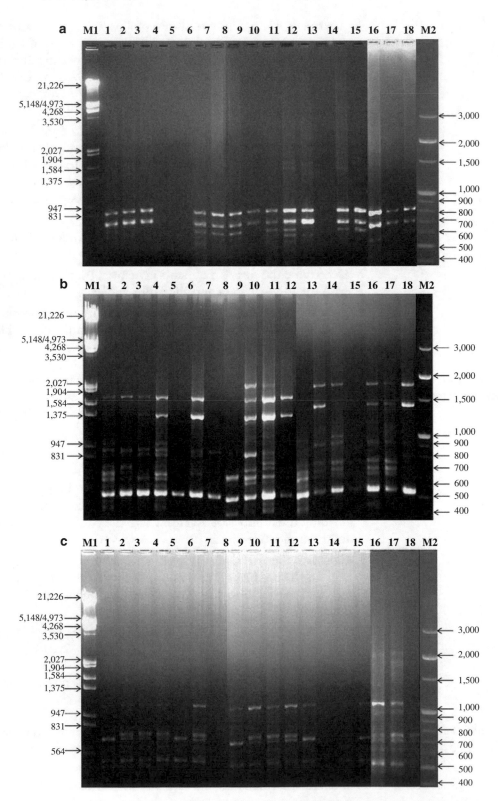

Fig. 1 Amplification products of ISSR primers (**a**) UBC-811, (**b**) UBC-825, and (**c**) UBC-842. Interpretation of figure is given in Table 2

Table 2
ISSR fingerprint data for the molecular characterization of *Aegle marmelos.* genotypes. The ISSR bands for the highlighted primers are shown in Fig. 1 (*see* Note 21)

S. No.	ISSR primers	No. of bands	Polymorphic bands	% polymorphism	PIC	EMR	MI
1.	UBC 811	7	5	71.4%	0.20	35	7.0
2.	UBC 815	2	2	100%	0.45	4	1.8
3.	UBC 824	3	2	66.6%	0.22	6	1.2
4.	UBC 825	11	10	90.9%	0.37	110	40.7
5.	UBC 836	8	6	75.0%	0.20	48	9.6
6.	UBC 840	5	5	100.0%	0.40	25	10
7.	UBC 842	6	5	83.3%	0.36	30	10.8
8.	UBC 859	3	2	66.6%	0.22	6	1.32
9.	UBC 888	14	14	100.0%	0.36	196	70.56
10.	UBC 889	14	14	100%	0.39	196	76.44
11.	UBC 890	13	10	76.9%	0.27	130	35.1
Total		86	75				
Mean/Avg. values		7.81 bands/primer	6.81	87.21%	0.31	71.45	24.04

and "monomorphic" if present in all individuals (Table 2). Monomorphic bands are not useful in genetic diversity analysis.

3. Create the 0 or 1 binary data matrix from binary data and use it to calculate the genetic distance/similarity (GS) using "SIMQUAL" subprogram of software NTSYS-PC (Numerical Taxonomy and Multivariate Analysis System for PC) [30] (*see* **Note 20**).

4. Construct a dendrogram (Fig. 2) based on SAHN cluster analysis [31], a subprogram of NTSYS-PC and Unweighted Pair Group Method with Arithmetic average (UPGMA) (*see* **Note 20**).

4 Notes

1. If the 1 M solution of Tris–HCl has a yellow color, discard it and obtain Tris of better quality. The pH of Tris solutions is temperature-dependent and decreases ~0.03 pH units for each 1 °C increase in temperature. For example, a 0.05 M solution has pH values of 9.5, 8.9 and 8.6 at 5 °C, 25 °C and 37 °C, respectively. To get Tris–HCl of pH 8.0, approximately 42 mL

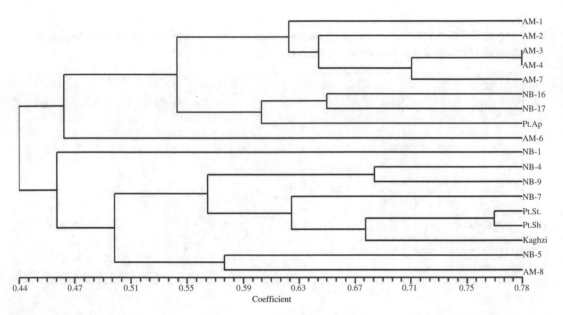

Fig. 2 Dendrogram based on genetic distance computed from ISSR using UPGMA algorithm in different geno-types of bael (*see* **Note 22**)

of 1 M HCl is required. So the volume of water added initially before pH adjustment with 1 M HCl should be kept small. While adjusting pH, 12 M HCl can be used first in order to reduce the difference between the starting pH (highly alkaline) and the required pH (8.0). Later on it is better to use 1 M HCl (lower ionic strength) to avoid a drastic and sudden decrease in pH below the required pH. pH adjustment should be done by adding HCl while the solution is stirred on magnetic stirrer and pH electrode is dipped in the solution (not too deep, as the electrode may get damaged by the rotating magnetic stir bar). Tris–HCl can be dissolved relatively easily by taking water at the bottom of the beaker. This allows the magnetic stir bar to go to work immediately.

The process of dissolution in water can also be expedited by using water preheated to 37 °C, which is then cooled before pH adjustment.

2. Dilution of stock solution to get the required final concentration is done according to the following formula: $C1 \times V1 = C2 \times V2$, where C1 = Conc. of stock solution; V1 = Volume of stock solution to be added to get required concentration; C2 = Final conc. required in a solution; V2 = Volume of solution to be prepared

1000 mM Tris–HCl $\times$ V1 = 100 $\times$ 10; Thus V1 = 1 mL

500 mM $Na_2EDTA \times$ V1 = 20 $\times$ 10; Thus, V1 = 0.4 mL

5000 mM NaCl $\times$ V1 = 2000 $\times$ 10; Thus V1 = 4 mL

3. pH adjustment should be done by adding NaOH pellets while the Na$_2$EDTA solution is stirred on magnetic stirrer and pH electrode is dipped in the solution (not too deep, as the electrode may get damaged by the rotating magnetic stir bar). It is advisable to adjust pH by adding NaOH pellets first in order to reduce the difference between the starting pH (acidic) and the required pH (8.0), followed by addition of 1 M NaOH in order to avoid a drastic and sudden increase in pH above the required pH. Similar to Tris–HCl, the volume of water added initially before pH adjustment with NaOH should be kept small. EDTA will not go completely into solution until the pH is adjusted to about 8.0.

4. 1.5% (w/v) PVP is added directly into the mortar during homogenization to precipitate and remove phenolic compounds from plant DNA extracts. 2.5% (v/v) β-ME is added as antioxidant/reducing agent, which breaks the intramolecular disulfide bonds in proteins, thereby increasing their denaturation. It helps in removing tannins and other polyphenols often present in the crude plant extract.

5. The pH of this buffer is not adjusted, but is ~8.5. 10× TAE can also be prepared as stock buffer instead of 50×. 1× TAE is used both as gel buffer and running/electrophoresis buffer. It not only establishes pH (8.0) but also provides ions to support conductivity. If water is used mistakenly instead of buffer, there will be essentially no migration of DNA in the gel. If concentrated buffer (10×) is used mistakenly, the gel may melt due to generation of enough heat. 1× TBE (Tris–borate–EDTA) buffer can also be used instead. The same buffer should be used for both preparation of the gel and running buffer. It should be prepared fresh to prevent nuclease contamination leading to DNA degradation.

6. EtBr is a fluorescent dye that intercalates between bases of nucleic acids and allows very convenient detection of DNA fragments in gels. The DNA fragments bound with EtBr appear as orange bands under UV light in gel documentation systems. Caution: Ethidium bromide is a mutagen and should be handled as a hazardous chemical. Wear gloves while handling. The gel containing EtBr should be discarded with utmost care.

7. EtBr at a concentration of 0.5 μg/mL may be added in agarose gel before solidification or mixed in sample to be loaded on gel or even the gel may be stained after completion of electrophoresis. The fluorescent yield of EtBr–DNA complex is greater than that of unbound dye, hence small amounts of DNA (~10 ng/band, 0.5 cm wide band) can be detected in the presence of EtBr (0.5 μg/mL) in gels.

8. As calculated in **Note 2**, 10 mL of extraction buffer can be prepared by adding 1 mL of 1 M Tris–HCl (pH 8.0), 0.4 mL of 0.5 M Na_2EDTA (pH 8.0), 4 mL of 5 M NaCl, and 2% (w/v) CTAB (solid) and remaining volume of sterile distilled water. Tris–HCl is used as buffer and thus gives the right pH (8.0) for DNA extraction. It also makes the cell membranes more permeable. Na_2EDTA is added as a chelating agent, which chelates divalent metal ions. It thus prevents the magnesium-mediated aggregation of nucleic acids with other biomolecules. The chelation of magnesium also inactivates nucleases. NaCl is added to provide positive ions (Na^+) to neutralize the negative charge of nucleic acids and thereby bring the nucleic acid molecules together. CTAB, a detergent, is added to disrupt plant cell wall and biological membranes and also to denature or inhibit proteins, including enzymes such as nucleases.

9. The extraction buffer used for homogenization is pre-warmed at 65 °C. This step is crucial for inactivation of plant nucleases.

10. Laying tubes sideways during isopropanol precipitation increases the efficiency of DNA precipitation by increasing the surface area of action.

11. This step is to wash any residual salt away from the pelleted DNA in 30% water that is mixed with ethanol, while preventing the DNA from dissolving itself. The presence of residual salts in the DNA sample may lead to smeared or shifted band patterns. Absolute ethanol is not used for washing. After washing, it is better to centrifuge at 10,000 rpm for 5 min, rather than simply decanting, as the DNA pellet may be lost if disturbed or loosened upon 70% ethanol addition. The DNA pellet should be air-dried to remove traces of ethanol. If the DNA sample contains residual ethanol after its preparation, it may float out of the well in the gel.

12. As a result of resonance, all the bases present in the nucleic acid absorb UV light. Nucleic acids are characterized by a maximum absorption of UV light at wavelength near 260 nm in a spectrophotometer. An optical density (OD; absorbance) of 1.0 corresponds to ~50 µg/mL for double stranded DNA. The DNA concentration is estimated in diluted (with water) samples using the equation:

$$µg\ DNA\ ml^{-1} = OD_{260} \times 50 \times dilution\ factor$$

13. The purity of DNA sample is assessed by calculating the ratio of ODs at 260 and 280 nm (OD_{260}/OD_{280}). For pure DNA sample, this ratio should be 1.8. Any deviation from this value indicates contamination; higher value indicates contamination with RNA and lower value indicates contamination with proteins.

14. BPB (3′,3″,5′,5″-tetrabromophenolsulfonphthalein) is used as tracking dye. It is a pH indicator dye and a color marker. At pH 8.0, it emits blue color and turns yellow at acidic pH; as such it can indicate any pH change in buffer during the electrophoretic run. BPB carries a slight negative charge at moderate pH and hence it migrates in the same direction as DNA in a gel. BPB migrates in the gel towards the anode at predictable rates under the influence of electric field and allows visual monitoring or how far the electrophoresis has proceeded. The rate of migration of BPB through agarose gels run in 1× TAE is approximately the same as linear double stranded DNA 300 bp in length. This relationship is not significantly affected by the concentration of agarose in the gel over the range of 0.5–1.5%. BPB thus moves at a speed faster than the smallest DNA fragment that can be separated on 0.5–1.5% gels (or even more) and hence can help in monitoring the progress of electrophoresis, consequently preventing the movement and loss of DNA in the electrophoresis buffer. BPB should not be used in excess, as it may cause DNA masking. 30% glycerol in the tracking dye makes the sample dense, ensuring that the DNA sinks evenly into the well, and to prevent diffusion or floating away of samples into the electrophoresis buffer. Sucrose or Ficoll at the same concentration can be used instead of glycerol.

15. Fresh electrophoresis buffers, poured gels, nuclease-free vials and tips should be used to minimize nuclease contamination of DNA solutions. When preparing agarose gels, the volume of water should be adjusted to accommodate for evaporation during boiling, otherwise the gel percentage will become higher and result in poor separation of larger DNA bands. The gel should be poured slowly to avoid formation of bubbles. Bubbles, if formed, can be removed with a pipette tip. Entrapment of any physical particle should also be prevented. Hence, pure sterile water, clean flasks and clean equipment should be used for preparation of gels. While inserting the comb into the gel, it should be kept vertical to the gel surface and stable during gel pouring and its solidification. The comb should be removed only after complete polymerization of the gel. 1× TAE buffer should be poured immediately after polymerization of gel and then the comb should be taken out. Improper handling of comb leads to formation of poorly formed or slanting wells. In such cases, DNA may remain in the wells during electrophoresis. The DNA sample solution should not contain any precipitate as it may prevent DNA migration. The sample or the ladder/marker volume should be large enough to fill 1/3rd of the total capacity of the well. Large wells should not be used with small sample volumes. The entire

gel should be completely immersed in the electrophoresis buffer during the run. Incompletely immersed gel during loading or electrophoresis and low sample volumes may lead to formation of curved bands. The gel and apparatus should be horizontally positioned during the run. The electrophoresis should not be performed at excessively high voltage. This may result in gel heating and DNA denaturation or may lead to atypical band patterns. Excessive electrophoresis run times or voltage may result in migration of small DNA fragments of the gel. Very short or slow electrophoresis may result in incompletely resolved bands. The gel should not be stored for long-term before taking a photograph, as this may cause diffusion of DNA fragments or lower the band intensity.

16. Standard ISSR primers (selected from literature) have been purchased from Eurofins Genomics Private India Limited. UBC stands for the University of British Columbia (The Michael Smith Laboratories, University of British Columbia, Vancouver, BC, Canada). (http://www.michaelsmith.ubc.ca/services/NAPS/Primer_Sets/Primers.pdf).

17. Optimization of the ISSR reaction parameters is mandatory; various concentrations (low and high) of PCR components, viz., DNA template, primer concentration (picomoles), dNTPs (mM), *Taq* DNA polymerase (units), and Mg^{2+}, are considered for the optimization of the ISSR-PCR reaction. The reaction is performed using the annealing temperature ranging between 50–55 °C. Moreover, care should be taken to add *Taq* DNA polymerase at the last, so as to avoid much reaction in the PCR tube. To avoid reaction to occur in the tube, hot start PCR can also be performed.

18. The *Eco*R I/*Hin*d III double digestion of λ DNA yields 13 discrete fragments of the sizes 21226, 5148, 4973, 4268, 3530, 2027, 1904, 1584, 1375, 947, 831, 564, 125 bp. The marker is ready-to-use. It is thus used to estimate DNA fragment size. The cohesive ends (12b *cos* site of bacteriophage) of fragments 21,226 bp and 3530 bp may anneal and form an additional band. These fragments may be separated by heating to 65 °C for 5 min and then cooling on ice for 3 min. DNA ladders are not heated before loading. The DNA ladder/DNA size marker are mixed with the same loading dye as that used for preparing DNA sample. Equal or very similar volumes of the sample DNA and the DNA ladder/DNA size marker should be loaded, otherwise atypical banding patterns may be obtained. The sample can be diluted with 1× loading dye.

19. The PIC value of a marker helps in detecting the polymorphism within a population and hence is a good index for genetic diversity evaluation. PIC index can be used to evaluate the level of gene variation, where

PIC >0.5 indicates that the locus is of high diversity,

PIC <0.25 indicates that the locus is of low diversity,

PIC between 0.25 and 0.5 indicates that the locus is of intermediate diversity.

The number of loci polymorphic in the germplasm set of interest, analyzed per experiment is called effective multiplex ratio (E or EMR). The average number of DNA fragments amplified or detected per genotype using a marker system is considered as multiplex ratio (n).

MI is interpreted as the average number of bands per experimental unit differentiating two randomly selected individuals. MI calculates the overall utility of a marker system.

High PIC value of a marker indicates it to be more informative than others. High EMR and MI values of a marker reflect the ability to simultaneously detect large number of bands.

20. Interpretation of Table 2. A total of 86 bands are obtained using the 16 primers with an average of 7.81 bands per primer. Such a high variation in the number of fragments produced by these arbitrary primers may be attributed to the differences in the binding sites throughout the genome of *Aegle marmelos*. Out of 86 bands, 75 are polymorphic and 11 are monomorphic. The number of bands produced per primer ranges from two (ISSR 815) to 14 (ISSR 888 and 889). The size of fragments ranges from 100 to 1500 bp. Eight unique bands are obtained with ISSR primers UBC 811, 836, 888, 890. A high level of polymorphism (100%) indicates the presence of high variability in bael with primers UBC 815, 840, 888, 890. PIC values range from 0.20 (UBC 811, 836) to 0.45 (UBC 815). The EMR values for ISSR range from 4 to 196 with an average of 71.45 per primer combination. The highest value of EMR (196) is scored with the primers UBC 888 and 889 and the lowest value (4) is scored with the primer UBC 815. The MI values for ISSR range from 1.2 to 76.44 with an average of 24.04 per primer combination. The highest value (76.44) is scored with the primer UBC 889 and the lowest value (1.2) is scored with the primer UBC 824.

21. SIMQUAL (Similarity for Qualitative data) computes various association coefficients for qualitative data like data with unordered states (e.g., Jaccard coefficient). SAHN (Sequential, Agglomerative, Hierarchical, Nested) performs the sequential, agglomerative, hierarchical, and nested clustering methods, which include a commonly used clustering method, e.g., UPGMA. This program can also find alternative trees in the case of ties in the input matrix.

22. Interpretation of Fig. 2. The accessions are clustered into two major clusters, each of them is further subdivided into subclusters

and subgroups showing varied degree of similarities. The two major clusters are genetically most diverse at a similarity value of 0.44. Highest similarity (0.78) is observed between AM-3 and AM-4. The genotype AM-7 merges with AM-3 and AM-4 at a similarity value of 0.71. These genotypes unite with AM-2 at a similarity value of 0.64. AM-1 clusters with these genotypes at 0.62. NB-16 shows similarity with NB-17 (0.65). Pant Aparna (PtAp) shows closeness with NB-16 and NB-17 having a similarity value of 0.60. These subgroups join the subcluster of one of the major clusters at a genetic distance of 0.55. Similarly, the varieties Pant Sujata (PtSt) and Pant Shivani (PtSh) show high similarity value of 0.76. Kaghzi merges with the above two at 0.68 similarity coefficient. The other genotypes which show similarity and cluster together are NB-4, NB-9and NB-5, AM-8.

Acknowledgements

University Grants Commission, India is gratefully acknowledged for the financial support of this investigation as M.A.N.F. to F.M.

References

1. Benharrat H, Veronesi C, Theodet C, Thalouam P (2002) *Orobanche* species and population discrimination using inter simple sequence repeat (ISSR). Weed Res 42: 470–474

2. Gonçalves LSA, Rodrigues R, Amaral Júnior AT, Karasawa M, Sudré CP (2008) Comparison of multivariate statistical algorithms to cluster tomato heirloom accessions. Genet Mol Res 7(4):1289–1297

3. Virk PS, Zhu J, Newburg HJ, Bryan GJ, Jeckson MT, Ford-Lloyd BV (2001) Effectiveness of different classes of molecular markers for classifying and revealing variation in rice germplasm. Euphytica 112:275–284

4. Fernandez ME, Figueiras AM, Benito C (2002) The use of ISSR and RAPD markers for detecting DNA polymorphisms, genotype identification and genetic diversity among barley cultivars with known origin. Theor Appl Gen 104: 845–851

5. Vicente MJ, Segura F, Aguado M, Migliaro D, Franco JA, Martínez-Sánchez JJ (2011) Genetic diversity of *Astragalus nitidiflorus*, a critically endangered endemic of SE Spain, and implications for its conservation. Biochem Syst Ecol 39:175–182

6. Rubio-Moraga A, Candel-Perez D, Lucas-Borja ME, Tiscar PA, Viñegla B, Linares JC, Gómez-Gómez L, Ahrazem O (2012) Genetic diversity of *Pinus nigra* Arn. Populations in southern Spain and northern Morocco revealed by inter-simple sequence repeat profiles. Int J Mol Sci 13:5645–5658

7. Chowdhury MA, Vandenberg B, Warkentin T (2002) Cultivar identification and genetic relationship among selected breeding lines and cultivars in chickpea (*Cicer arietinum* L.) Euphytica 127:317–325

8. Mattioni C, Casasoli M, Gonzalez M, Ipinza R, Villani F (2002) Comparison of ISSR and RAPD markers to characterize three Chilean Nothofagus species. Theor Appl Genet 104: 1064–1070

9. Goulão L, Oliveira CM (2001) Molecular characterization of cultivars of apple (*Malus domestica* Borkh.) using microsatellite (SSR and ISSR) markers. Euphytica 122:81–89

10. Tian HL, Xue JH, Wen J, Mitchell G, Zhou S-L (2008) Genetic diversity and relationships of lotus (*Nelumbo*) cultivars based on allozyme and ISSR markers. Sci Hortic 116(4):421–429

11. Reddy MP, Sarla N, Siddiq EA (2002) Inter simple sequence repeat (ISSR) polymorphism

and its application in plant breeding. Euphytica 128:9–17

12. Triest L (2007) Molecular ecology and biogeography of mangrove trees towards conceptual insights on gene flow and barriers: a review. Aquat Bot 89:138–154

13. Wolfe AD, Liston A (1998) Contributions of PCR-based methods to plant systematics and evolutionary biology. In: Soltis DE, Soltis PS, Oyle JJ (eds) Plant Mol sys II. Kluwer, Boston

14. Saki S, Bagheri H, Deljou A, Zeinalabedini M (2016) Evaluation of genetic diversity amongst *Descurainia sophia* L. genotypes by inter-simple sequence repeat (ISSR) marker. Physiol Mol Biol Plants 22(1):97–105

15. Liu LW, Zhao LP, Gong YQ, Wang MX, Chen LM, Yang JL, Wang Y, Yu FM, Wang LZ (2008) DNA fingerprinting and genetic diversity analysis of late-bolting radish cultivars with RAPD, ISSR and SRAP markers. Sci Hortic 116:240–247

16. Wan YT, Xin-Xiang A, Fan CZ, Xu FR, Yu TQ, Tang CF, Dai LY (2008) ISSR analysis on genetic diversity of the 34 populations of *Oryza meyeriana* distributing in Yunnan province, China. Rice Sci 15(1):13–20

17. Huang Y, Ji KS, Jiang ZH (2008) Genetic structure of *Buxus sinica* Var. Parvifolia, a rare and endangered plant. Sci Hortic 116: 324–329

18. Thangjam R (2014) Inter-simple sequence repeat (ISSR) marker analysis in *Parkia timoriana* (DC.) Merr. Populations from Northeast India. Appl Biochem Biotechnol 172:1727–1734

19. Mishra P, Kumar LD, Kumar A, Gokul S, Ravikumar K, Shukla AK, Sundaresan V (2015) Population dynamics and conservation implications of *Decalepis arayalpathra* (J. Joseph and V. Chandras.) Venter, a steno endemic species of Western Ghats, India. Appl Biochem Biotechnol 176:1413–1430

20. Prevost A, Wilkinson MJ (1999) A new system of comparing PCR primers applied to ISSR fingerprinting of potato cultivars. Theor Appl Genet 1:107–112

21. Casasoli M, Mattioni C, Cherubini M, Villani F (2001) A genetic linkage map of European chestnut (*Castanea sative* mill.) based on RAPD, ISSR and isozyme markers. Theor Appl Genet 102:1190–1199

22. Arnau G, Lallemand J, Bourgoin M (2002) Fast and reliable strawberry cultivar identification using inter simple sequence repeat (ISSR) amplification. Euphytica 129:69–79

23. Archak S, Gaikwad AB, Gautam D, Rao EVVB, Swamy KRM, Karihaloo JL (2003) DNA fingerprinting of Indian cashew (*Anacardium occidentale* L.) varieties using RAPD and ISSR techniques. Euphytica 130:397–404

24. Vijayan K, Chatterjee SN (2003) ISSR profiling of Indian cultivars of mulberry (Morus spp.) and its relevance to breeding programs. Euphytica 131:53–63

25. Gardner N, Hokanson SC (2005) Intersimple sequence repeat fingerprinting and genetic variation of clematis cultivars and commercial germplasm. HortSci 40:1982–1987

26. Sambrook J, Russell DW (2001) Molecular cloning: a laboratory manual. CSHL Press, New York

27. Ghislain M, Zhang D, Fajardo D, Huaman Z, Hijmans RJ (1999) Marker-assisted sampling of the cultivated Andean potato *Solanum phureja* collection using RAPD markers. Genet Resour Crop Evol 46:547–555

28. Tiwari S, Tiwari R, Srivastava R, Kumari P, Singh SK (2013) Molecular genetic analysis of *Eucalyptus tereticornis* by using RAPD markers. Eur J Exp Biol 3(6):103–110

29. Varshney RKG, Sorrells ME (2005) Genomics-assisted breeding for crop improvement. Trends Plant Sci 10:621–630

30. Rohlf FJ (1998) NTSYS-PC: Numerical Taxonomy and Multivariate Analysis System, Version 2.0. Exeter Software, New York, USA.

31. Sneath PHA, Sokal RR (1973) Numerical taxonomy. Freeman, San Francisco, CA

PCR in the Analysis of Clinical Samples: Prenatal and Postnatal Diagnosis of Inborn Errors of Metabolism

Laura Vilarinho and Célia Nogueira

Abstract

Inborn errors of metabolism (IEMs) are individually rare but collectively common. As more and more genes are cloned and specific disease-causing mutations are identified, the diagnosis of IEMs is becoming increasingly confirmed by mutation analysis. Diagnosis is important not only for treatment and prognosis but also for genetic counselling and prenatal diagnosis in subsequent pregnancies. A wide range of molecular methods is available for the identification of mutations and other DNA variants, most of which are based on the Polymerase Chain Reaction (PCR). In this chapter, we focus on PCR-based methods for the detection of point mutations or small deletions/insertions as these are the most frequent causes of IEMs.

Key words PCR, IEM, Prenatal molecular diagnosis, Postnatal molecular diagnosis

1 Introduction

Inborn errors of metabolism (IEMs) are a heterogeneous group of disorders with variable clinical manifestations presenting mainly in the pediatric population [1]. In most of the disorders, problems arise due to accumulation of substances which are toxic or interfere with normal function, or to the effects of reduced ability to synthesize essential compounds. IEMs are now often referred to as congenital metabolic diseases or inherited metabolic diseases, and these terms are considered synonymous. Close to 10% of diseases among hospitalized children have been attributed to Mendelian traits inherited as single gene defects, not a surprising figure considering that approximately 1000 IEMs have been identified to date [2]. Although individual IEMs are rare, collectively they represent a large and diverse class of genetic conditions, with new disorders and disease mechanisms being described regularly [3]. In recent years, dramatic advances in the applications of molecular genetic techniques have made it possible to identify molecular defects in human that account for IEMs [4]. Thus, molecular genetic techniques have rapidly expanded our ability to diagnose disease.

Lucília Domingues (ed.), *PCR: Methods and Protocols*, Methods in Molecular Biology, vol. 1620,
DOI 10.1007/978-1-4939-7060-5_15, © Springer Science+Business Media LLC 2017

DNA diagnosis includes prenatal and presymptomatic diagnosis as well as identification of carriers of many IEMs [5]. Disorders that are always caused by a specific mutation can be diagnosed by direct detection of the "common" mutation. DNA amplification by the PCR has made possible the rapid identification of DNA sequence variations. After PCR, mutations can be detected by DNA sequencing analysis [6]. These methods have made DNA diagnosis faster, easier, and cheaper. Early recognition and implementation of adequate therapeutic measures are of the highest importance in minimizing morbidity and improving clinical outcome [7]. In severe cases, a molecular prenatal diagnosis can be offer if the index case was previously studied and the disease causing mutation was identified.

2 Materials

2.1 DNA Extraction

Various commercially available DNA extraction kits and systems are becoming increasingly popular because of their ease of use, limited labor, and ability to consistently produce high-quality DNA. Because of proprietary considerations of the manufacturer, the composition of some components in these kits is not revealed to the user. These kits contain a stable proteinase K solution, unidentified non-phenol–chloroform buffers, a silica-gel membrane spin-column or magnetic-particles to isolate and purify high-quality DNA for PCR.

2.2 PCR Requirements

2.2.1 Oligonucleotide Primers

Oligonucleotides are widely available and there are many companies (such as Thermo scientific, Invitrogen, MWG Biotech, Applied Biosystems, Sigma Genosys, among others) that offer low-cost custom synthesis and purification of your primer sequences within a few days of ordering. For most PCRs you will need two primers of different sequence that anneal to complementary strands of the template DNA. When you know the DNA sequence of your template it is quite easy to design appropriate primers to amplify any segment that you require [8]. There are several computer programs that can be used to assist primer design.

2.2.2 PCR Premixes

Several commercial PCR premixes, such as ImmoMix Red (Bioline), AmpliTaq Gold (Applied Biosystems), TaKaRa Taq DNA Polymerase (TaKaRa-Clontech), among others are available. These premixes contain buffer, dNTPs and Taq DNA polymerase as a premixed reagent at a concentration that allows addition of template DNA and primers to produce the final reaction volume. In some cases the buffers contain no magnesium, allowing optimization experiments to be undertaken by addition of magnesium stocks. Clearly, the use of premixes is highly advantageous for high-throughput screening or template preparation applications [9]. Before taking an aliquot to add to your PCRs always remember to thoroughly thaw out and mix the solutions.

2.2.3 PCR Cleanup and Sequencing

1. ExoSAP-IT reagent treats PCR products ranging in size from less than 100 bp to over 20 kb with absolutely no sample loss by removing unused primers and nucleotides. ExoSAP-IT PCR Product clean up is active in commonly used PCR buffers, so no buffer exchange is required.

2. The BigDye™ Terminator v1.1 Cycle Sequencing Kit provides premixed reagents for Sanger sequencing reactions [10]. The kit includes BigDye™ Terminator v 1.1, [5×] Sequencing Buffer, which is specifically optimized for use with the BigDye™ Ready Reaction mixes.

3. DyeEx Kits provide either spin columns or 96-well plates and use gel-filtration technology to remove unincorporated dye terminators from sequencing reactions. Sequencing reactions are loaded onto the prehydrated gel-filtration material. After a short centrifugation step, the reactions are ready to be loaded onto a capillary sequencer. Unincorporated dye terminators are retained in the gel matrix.

2.3 Agarose Gel Electrophoresis

1. Agarose gels are prepared using a w/v percentage solution and should be diluted in running buffer. The concentration of agarose in a gel will depend on the sizes of the DNA fragments to be separated, with most gels ranging between 0.5 and 2%.

2. The most common gel running buffers are [1×] TAE (40 mM Tris–acetate, 1 mM EDTA) and [1×] TBE (45 mM Tris–borate, 1 mM EDTA).

3. [1×] GelRed (Biotium).

3 Methods

3.1 DNA Extraction

DNA from almost any source has been successfully used in PCR. However, for **prenatal diagnosis of IEMs** we usually isolated DNA directly from amniotic fluid or chorionic villi, and/or cultured amniocytes or chorionic villi cells. For **Postnatal diagnosis of IEMs** DNA is usually extracted from total blood, Guthrie cards, and/or tissues (fibroblasts, muscle, or liver).

In general, PCR is almost certain to work best with the cleanest, purest DNA sample you can prepare.

Most DNA preparation protocols outline the steps involved in four major phases of DNA isolation:

1. The first phase contains preparatory procedures and the solutions used during these steps.

2. The second phase of the protocols is the actual isolation steps required to separate the nucleic acids from the rest of the cellular proteins.

3. The third phase in these preparations includes steps required to purify nucleic acids from impurities that will interfere with subsequent enzymatic manipulations.

4. In the fourth phase DNA is resuspended in nuclease-free water or in TE Buffer Solution.

3.2 Quantification of DNA

1. DNA concentration can be determined by measuring the absorbance at 260 nm ($A260$) in a spectrophotometer NanoDrop 2000c (Thermo Scientific) [11].

2. Clean the upper and lower optical surfaces of the microvolume spectrophotometer sample retention system by pipetting 2–3 μL of clean deionized water onto the lower optical surface.

3. Close the lever arm, ensuring that the upper pedestal comes in contact with the deionized water. Lift the lever arm and wipe off both optical surfaces with a clean, dry, lint-free lab wipe.

4. Open the NanoDrop software and select the Nucleic Acid application. Use a small-volume, calibrated pipettor to perform a blank measurement by dispensing 1 μL of buffer onto the lower optical surface. Lower the lever arm and select "Blank" in the Nucleic Acid application.

5. Once the blank measurement is complete, clean both optical surfaces with a clean, dry, lint-free lab wipe.

6. Choose the appropriate constant for the sample that is to be measured: dsDNA; DNA-50 (an absorbance of 1 unit at 260 nm corresponds to 50 μg genomic DNA per mL ($A260 = 1$ for 50 μg/mL)).

7. Dispense 1 μL of nucleic acid sample onto the lower optical pedestal and close the lever arm. Because the measurement is volume independent, the sample only needs to bridge the gap between the two optical surfaces for a measurement to be made.

8. Select "Measure" in the application software. The software will automatically calculate the nucleic acid concentration and purity ratios. Following sample measurement, review the spectral output.

9. The software will automatically calculate the nucleic acid concentration and purity ratios. Following sample measurement, review the spectral image to assess sample quality.

10. The ratio of the readings at 260 and 280 nm ($A260/A280$) provides an estimate of DNA purity with respect to contaminants that absorb UV light, such as protein.

11. Pure DNA should present an $A260/A280$ ratio between 1.8 and 2.0.

12. The ratios obtained <1.8 resulting in protein contamination and the >2.0 are attributed to RNA contamination.

13. Depending on the DNA isolation method used, RNA will be copurified with genomic DNA. RNA may inhibit some downstream applications, but it will not inhibit PCR reaction. Treatment with RNase A will remove contaminating RNA.

3.3 Primers Design

Primers can be designed using several computer programs, such as: Primer3 (http://bioinfo.ut.ee/primer3-0.4.0/), in order to flank the coding sequences and exon–intron junctions of the genes, and FastPCR   program   (http://www.biocenter.helsinki.fi/bi/Programs/fastpcr.htm), to predict the formation of dimers and to evaluate the quality of the designed primers [12].

However, in practice many technicians still design primers by following some simple rules:

1. A primer should be 16–30 nucleotides long, which provides good specificity for a unique target sequence, even with a starting template as complex as human genomic DNA.

2. Primers should contain approximately equal numbers of each nucleotide.

3. Primers should not include stretches of polybase sequences (e.g., poly (dG)) or repeating motifs, as these can hybridize inappropriately to the template.

4. Primer pairs should have compatible melting temperatures (within 5 °C) and contain approximately 50% GC content. High GC content results in the formation of stable imperfect hybrids, while high AT content depresses the melting temperature of perfectly matched hybrids. If possible, the 3′ end of the primer should be rich in GC bases (GC clamp) to enhance annealing of the end that will be extended, but should not exceed 3 Gs or Cs.

5. Primers should not be able to form secondary structures due to internal complementarity.

6. Primers should not contain sequences at the 3′-ends that will allow base pairing with itself or any other primer that it may be coupled with in a PCR, otherwise this can lead to the formation of primer dimers (*see* **Note 1**).

7. The first three nucleotides at the 3′-end should perfectly match the template with complementarity extending to about 20 bp with a few mismatched bases, because this is the end of the primer that is extended by the DNA polymerase and is therefore most important for ensuring the specificity of annealing to the correct target sequence.

8. The 5′-end of the primer is less important in determining specificity of annealing to the target sequence and this means it is possible to alter the sequence in some desirable manner to facilitate subsequent cloning, manipulation, mutagenesis, recombination, or expression of the PCR product.

3.3.1 Universal-Tailed Primers (M13 Primers)

Primers could be tagged with generic sequencing universal sequences such as M13 Forward Primer (5′-TGTAAAACGACG GCCAGT-3′) and M13 Reverse Primer (5′-CAGGAAACAGCTA TGACC-3′). M13-tailed PCR primers contain at their 5′ ends a short string of nucleotides derived from M13 DNA. These short nucleotide sequences are viral in origin and therefore not represented in the human template. These primers are used for the sequencing reaction, independent of the PCR product to sequence. This procedure is less complicated, requires less preparation time, and minimizes sources of error (*see* **Note 2**).

3.4 PCR Procedure

1. Remove the following reagents from −20 °C freezer and keep on ice throughout this procedure.

2. Mix gently by vortex and briefly centrifuge to collect all components to the bottom of the tube.

3. Label microcentrifuge tubes and add components as indicated as follows:

Component	Volume (μL)	Final concentration
[2×] PCR premixes	6.25	[1×]
10 μM forward primer	0.5	0.4 μM (0.05–1 μM)
10 μM reverse primer	0.5	0.4 μM (0.05–1 μM)
Template DNA	Variable	<1000 ng
Nuclease-free water	to 12.5	

4. Amplification conditions:

			Desnaturation	Anneling		Extention		
Temperature	95 °C	94 °C		50–65 °C[a]	72 °C	72 °C	4 °C	
Time	10 min	1 min		1 min	1 min	5 min	Until ready to purify	
				35 cycles				

aFor each fragment the PCR conditions are optimized using annealing temperature gradients [13]

The annealing temperatures used for optimized each fragment are: 50 °C; 55 °C; 60 °C, and 65 °C, as indicated in Fig. 1.

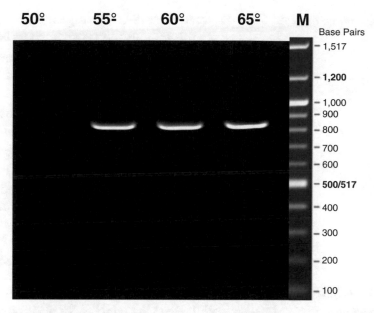

Fig. 1 Agarose gel electrophoresis of PCR products obtained from an annealing temperature gradient (50–65 °C). At 50 °C no amplification was found for this fragment. The DNA size marker (M) is a commercial 100-bp DNA ladder

3.5 Agarose Gel Electrophoresis

1. Weigh out the appropriate amount of agarose into an Erlenmeyer flask (*see* **Note 3**).

2. Add Running buffer.

3. Melt the agarose–buffer mixture in a microwave. At 30-s intervals, remove the flask and swirl the contents to mix well. Repeat until the agarose is completely dissolved.

4. Add [1×] GelRed.

5. Place the gel tray into the casting apparatus and put an appropriate comb into the gel mold to create the wells.

6. Allow the agarose to set at room temperature.

7. Remove the comb and place the gel in the gel box.

8. Add loading dye to the DNA samples to be separated, in the case of uncolored PCR mixes. An appropriate DNA size marker should always be loaded along with experimental samples.

9. Program the power supply to desired voltage (1–5 V/cm between electrodes) and run for 15–20 min.

10. The PCR amplification product can be analyzed in 2% agarose in [1×] TAE buffer with GelRed and visualized in a Gel Doc System.

11. Agarose gel analysis enables quick and easy quantification of DNA, especially for small DNA fragments (such as PCR products). The amount of sample DNA loaded can be estimated by comparison of the band intensity with the standards used.

3.6 PCR Clean Up

1. Remove ExoSAP-IT reagent from −20 °C freezer and keep on ice throughout this procedure (*see* **Note 4**).

2. Mix 2 μL of a post-PCR reaction product with 1 μL of ExoSAP-IT reagent for a combined 3 μL reaction volume (*see* **Note 5**).

3. Incubate at 37 °C for 15 min in the thermocycler to degrade remaining primers and nucleotides.

4. Incubate at 80 °C for 15 min in the thermocycler to inactivate ExoSAP-IT reagent.

5. The treated PCR product is now ready for use in DNA sequencing, single nucleotide polymorphism (SNP) analyses, or other primer-extension applications that require DNA to be free of excess primers and nucleotides (*see* **Note 3**).

6. Treated PCR products may be stored at −20 °C until required.

3.7 Sequencing Reaction

1. Completely thaw the contents of the BigDye™ Terminator v1.1 Cycle Sequencing Kit and your primers and store on ice.

2. Vortex the tubes for 2–3 s, then centrifuge briefly (2–3 s) with a benchtop microcentrifuge to collect contents at the bottom of the tubes.

3. Label microcentrifuge tubes "forward" and "reverse" and add components as indicated:

Component	Quantity per reaction (μL)
BigDye™ Terminator 3.1 Ready Reaction Mix	0.5
M13 forward or M13 reverse primer or primer forward or primer reverse (3.2 μM)	0.5
[5×] sequencing buffer	0.5
Template	3.0
Deionized water (RNase/DNase-free)	5.5
Total volume	10.0

4. Vortex the plate for 2–3 s, then centrifuge briefly in a swinging bucket centrifuge to collect contents to the bottom of the wells (5–10 s) at $1000 \times g$.

5. Place the tubes or plate(s) in a thermal cycler and set the volume.

6. Perform cycle sequencing as indicated:

	Desnaturation	Anneling	Extention	
Temperature	94 °C	94 °C	50 °C	62 °C	4 °C
Time	4 min	10 s	5 s	4 min	Until ready to purify
		25 cycles		

3.8 Sequencing Reaction Cleanup

1. Take the DyeEx 96 plate out of the bag, and remove the tape sheets first from the bottom and then from the top of the DyeEx 96 plate. When handling the DyeEx 96 plate, ensure that it remains horizontal.

2. Place the DyeEx 96 plate on the top of the collection plate (provided; reusable) and centrifuge for 1 min at the calculated speed. Discard the flow-through. Always use the waste collection plates provided with the DyeEx 96 Kit.

3. Place the DyeEx 96 plate on top of the collection plate, add 300 μL deionized water to each well, and centrifuge for 3 min at the calculated speed (*see* **Note 6**).

4. Carefully place the DyeEx 96 plate on an appropriate elution plate with a suitable adapter.

5. Slowly apply the 10 μL of sequencing samples to the gel bed of each well (*see* **Notes 7** and **8**).

6. Centrifuge for 3 min at the calculated speed. The eluate contains the purified sequencing reaction (*see* **Note 9**).

3.9 Automated DNA Sequencing

1. The sequencing runs were performed in Genetic Analyzers, which use capillary electrophoresis.

2. The products of the cycle sequencing reaction are injected electrokinetically into capillaries filled with polymer.

3. High voltage is applied so that the negatively charged DNA fragments move through the polymer in the capillaries toward the positive electrode.

4. Capillary electrophoresis can resolve DNA molecules that differ in molecular weight by only one nucleotide.

5. Shortly before reaching the positive electrode, the fluorescently labeled DNA fragments, separated by size, move through the path of a laser beam.

6. An optical detection device on genetic analyzers detects the fluorescence signal.

7. The data is then displayed as an electropherogram (Fig. 2).

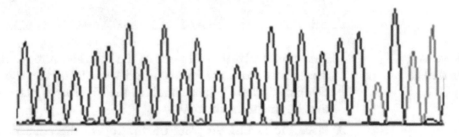

Fig. 2 Electropherogram showing a sequence data

3.10 Bioinformatic Analysis Tools

1. The sequences obtained can be compared to the reference of the respective gene using the Ensemble Genome Browser (http://www.ensemble.org) or using the SeqScape® Software (Applied Biosystems), and the variants found searched for in HGMD® Professional database (https://portal.biobase-international.com/hgmd/pro/start.php?).

2. Using the Clustal W program (http://www.ebi.ac.uk/Tools/msa/clustalw2/), multialignments should be checked to determine the degree of evolutionary conservation of the amino acid residue involved [14].

3. Furthermore, the amino acid substitution can be analysed by PolyPhen (PolyPhen: Prediction of functional effect of human nsSNPs. http://genetics.bwh.harvard.edu/pph/), to predict if an amino acid change is likely to be deleterious to protein function. Profile scores of >2.0 indicate that the variant is probably damaging to protein function whereas scores of 1.5–2.0 are possibly damaging, and scores of <1.5 are indicative of a benign variant.

4. To confirm results from PolyPhen, one should also use (a) the MutPred Server (http://mutpred.mutdb.org/), which generates a probability for a given mutation to be deleterious to protein function (scores that range from 0 to 1), and the likelihood of it being pathogenic (scores over 0.5) and (b) SIFT (http://sift.jcvi.org/), which yields a score of ≤0.05 if the amino acid substitution is predicted to be damaging and >0.05 if it is predicted to be tolerated.

3.11 Genotype Confirmation

1. Mutations must be confirmed by resequencing a second amplicon with both sense and antisense primers, after performing a new PCR amplification. Segregation of the mutations in available

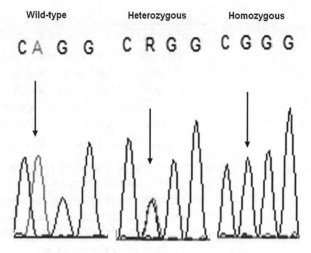

Fig. 3 *LPIN1* splicing mutation (c.2142-2A > G) found in heterozygous and homozygous state, in parents and in the index case, respectively, compared to the wild-type sequence

family members (Fig. 3) and frequency of novel mutations in a panel of 200 ethnically matched healthy chromosomes could be performed by direct sequencing.

2. For prenatal diagnosis, after exclusion of maternal contamination, the results obtained directly from amniotic fluid or chorionic villi should be confirmed in cultured amniocytes or chorionic villi cells.

4   Notes

1. A primer dimer is the product of primer extension either on itself or on the other primer in the PCR. Since the primer dimer product contains one or both primer sequences and their complementary sequences they provide an excellent template for further amplifications. Primer dimers can dominate the PCR and remove primer from the real target on the template DNA, because small products are copied more efficiently.

2. Primers could also be designed without generic universal sequences, but for the sequencing reaction each primer, forward and reverse, should be used in the respective PCR product.

3. The volume of the buffer should not be greater than one-third of the capacity of the flask.

4. Store ExoSAP-IT reagent in a non-frost-free freezer.

5. When treating PCR product volumes greater than 2 μL, simply increase the amount of ExoSAP-IT reagent proportionally.

6. The use of DyeEx 96 plates requires a suitable centrifuge capable of centrifuging microplates of 4.5 cm total height. All centrifugation steps are carried out at $1000 \times g$ at room temperature (15 °C–25 °C). The appropriate speed can be calculated: rpm = $1000 \times \sqrt{1000/1.12} \, r$ (r = radius of rotor in mm).

7. A multichannel pipet facilitates handling of sequencing samples.

8. Pipet the sequencing reaction directly onto the center of the gel-bed surface, without touching the reaction mixture or the pipet tip to the sides of the wells. The samples should be pipetted slowly so that they are absorbed into the gel and do not flow down the sides of the gel bed. Avoid touching the gel-bed surface with the pipet tip.

9. If using a formamide loading buffer, dry the samples and proceed according to the instructions provided with the DNA sequencer.

References

1. Mak CM, Lee HC, Chan AY et al (2013) Inborn errors of metabolism and expanded newborn screening: review and update. Crit Rev Clin Lab Sci 50:142–162

2. Ezgu F (2016) Inborn errors of metabolism. Adv Clin Chem 73:195–250

3. Vernon HJ (2015) Inborn errors of metabolism: advances in diagnosis and therapy. JAMA Pediatr 169:778–782

4. Rinaldo P, Hahn S, Matern D (2004) Clinical biochemical genetics in the twenty-first century. Acta Paediatr Suppl 93:22–26

5. Kamboj M (2008) Clinical approach to the diagnoses of inborn errors of metabolism. Pediatr Clin N Am 55:1113–1127

6. Sharma S, Kumar P, Agarwal R et al (2008) Approach to inborn errors of metabolism presenting in the neonate. Indian J Pediatr 75:271–276

7. Mahdieh N, Rabbani B (2013) An overview of mutation detection methods in genetic disorders. Iran J Pediatr 23:375–388

8. Van Pelt-Verkuil E, van Belkum A, Hays J (eds) (2008) Principles and technical aspects of PCR amplification. Springer Science, Berlin/Heidelberg, Germany

9. McPherson MJ, Moller SG (eds) (2007) PCR second edition. The BASICS. Garland Science, New York

10. Hernandez-Rodriguez P, Gomez A (2012) Polymerase chain reaction: types, utilities and limitations. In: Hernandez-Rodriguez P (ed) Polymerase chain reaction. InTech

11. Desjardins PR, Conklin DS (2011) Microvolume quantitation of nucleic acids. Curr Protoc Mol Biol 3:3J

12. Kalendar R, Lee D, Schulman AH (2014) FastPCR software for PCR, in silico PCR, and oligonucleotide assembly and analysis. Methods Mol Biol 1116:271–302

13. Munshi A (ed) (2012) DNA sequencing–methods and applications. InTech, Rijeka

14. Thompson JD, Higgins DG, Gibson TJ (1994) CLUSTAL W: improving the sensitivity of progressive multiple sequence alignment through sequence weighting, position-specific gap penalties and weight matrix choice. Nucleic Acids Res 22:4673–4680

Chapter 16

Harnessing the Power of PCR Molecular Fingerprinting Methods and Next Generation Sequencing for Understanding Structure and Function in Microbial Communities

Sujal Phadke*, Andreia Filipa Salvador*, Joana Isabel Alves*, Orianna Bretschger, Maria Madalena Alves, and Maria Alcina Pereira

Abstract

Polymerase chain reaction (PCR) is central to methods in molecular ecology. Here, we describe PCR-dependent approaches useful for investigating microbial diversity and its function in various natural, human-associated, and built environment ecosystems. Protocols routinely used for DNA extraction, purification, cloning, and sequencing are included along with various resources for the statistical analysis following gel electrophoresis-based methods (DGGE) and sequencing. We also provide insights into eukaryotic microbiome analysis, sample preservation techniques, PCR troubleshooting, DNA quantification methods, and commonly used ordination techniques.

Key words Polymerase chain reaction (PCR), Microbiome, Canonical correspondence analysis (CCA), Restriction fragment length polymorphism (T-RFLP), Denaturing gradient gel electrophoresis (DGGE), Next generation sequencing (NGS)

1 Introduction

Linking genomics to physiology and ecology through molecular biology methods has opened a wide and exciting perspective on the study of the vast number (80–99 %) of microbial species that have not yet been cultivated or are very difficult to cultivate in laboratory [1]. As a result, culture-independent methods are nowadays extensively used and are in continuous development for the study of microbial communities in their natural habitats, and also in engineered systems. Most of microbial ecology methods are PCR (polymerase chain reaction)-dependent, which means that

*These authors have contributed equally to this work.

Lucília Domingues (ed.), *PCR: Methods and Protocols*, Methods in Molecular Biology, vol. 1620,
DOI 10.1007/978-1-4939-7060-5_16, © Springer Science+Business Media LLC 2017

DNA needs to be amplified prior to further analysis. Depending upon the research question being addressed, the molecular targets for PCR include either rRNA genes that aid in phylogenetic inference or protein-coding genes that are important for specific cellular functions. For instance, taxonomic information is drawn from the 16S rRNA or 18S rRNA gene sequences obtained from the community, whereas protein-coding genes can be used to analyze specific metabolic capacities and diversification of the microbial diversity.

Many PCR-dependent molecular fingerprinting methods such as DGGE (denaturing gradient gel electrophoresis), T-RFLP (terminal restriction fragment length polymorphism), ARDRA (amplified ribosomal DNA restriction analysis), and ARISA (automated ribosomal interspace analysis) have been successfully used to evaluate the diversity of complex microbial communities. These methodologies can be further complemented with other approaches, such as cloning and sequencing, in order to obtain information about both the overall microbial diversity (species richness) and taxonomic identification. The main advantages of these techniques are that they are relatively fast and easy to perform, allow for the direct comparison of several samples at the same time and are culture-independent. Patterns obtained with fingerprinting methods can be compared using clustering algorithms that reveal phylogenetic relationships based on dendrograms and distance matrices, and allow determining similarity between microbial communities in distinct samples.

The DGGE fingerprinting technique involves amplification of a DNA fragment corresponding to a gene under study. Amplicons are then separated in a polyacrylamide gel containing a gradient of denaturing agents such as urea and formamide. The technique is based in the rationale that DNA fragments with different nucleotide sequences denature at different concentrations of urea and formamide, and at that point, stop migrating in the electrophoresis gel. Amplicons are generated using one of the primers (forward or reverse) carrying a "GC-clamp," which is a ~40 bp-long stretch of random sequence containing guanine and cytosine nucleotides. The GC-clamp avoids complete dissociation of double-stranded DNA and allows one to separate and distinguish DNA fragments from different origins in the same sample.

The T-RFLP technique involves amplification of a mixture of DNA fragments followed by their digestion with restriction enzymes that cut the fragments at specific DNA sequence motifs. The fragments treated with such enzymes are then separated using gel electrophoresis. The rationale is that differences in nucleotide sequences between species are revealed by the pattern in which the restriction enzymes cut the DNA fragments. The initial PCR step in T-RFLP may involve one or both primers being fluorescent-labeled. Fragments are separated either by capillary or polyacrylamide gel electrophoresis. Only the terminal fragments that contain the labeled end are detected and the remaining ignored. In RFLP (restriction fragment length polymorphism) and ARDRA the principle is similar, but all restriction fragments generated can be visualized. The peaks present in the T-RFLP electropherograms can be compared to peaks corresponding to specific microorganisms in databases (e.g., SILVA, GreenGenes, or NCBI) in order to provide taxonomic identification. Alternatively, taxonomy can be inferred by complementing T-RFLP or DGGE profiles with clone libraries.

In ARISA fingerprinting, the first step is to amplify the ITS (Internal Transcribed Spacer) region located between the 16S rRNA and 23S rRNA genes in prokaryotes (*Bacteria* and *Archaea*). The ITS is a noncoding region, highly variable in sequence length and nucleotide composition. When using fluorescent primers, results are visualized in an electropherogram. The peaks obtained can be correlated to the abundance of a given spacer region within a sample. Different electropherograms are compared to estimate microbial communities' diversity and similarity among distinct samples.

In this chapter, detailed procedures for these PCR-dependent methods will be presented (Fig. 1).

Most studies involve analysis of the prokaryotic (*Bacteria* and *Archaea*) microbiome. However, exploring microbial eukaryotic diversity is becoming increasingly necessary and customary to obtain the full picture of the structural and functional composition of the community. Most commercially available DNA extraction

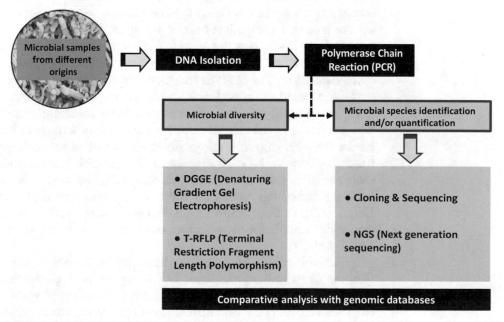

Fig. 1 PCR-dependent methods covered in the chapter

kits commonly used in microbiome analysis also yield eukaryotic DNA. For the analysis of eukaryotic diversity, PCR is usually performed using universal primers targeting the gene coding for 18S rRNA. However, different regions of the 18S rRNA gene may yield distinct diversity profiles and it is recommended that multiple 18S rRNA regions be used for a comprehensive analysis [2, 3]. In addition, primers for specific groups of microbial eukaryotes such as stramenopiles, ciliates, and fungi are also available [4–7]. The most commonly used universal 18S rRNA primers (5′-3′) are given below. V4 region of the 18S rRNA gene: forward primer (Fwd1): CCAGCASCYGCGGTAATTCC and reverse primer (Rev3): ACTTTCGTTCTTGATYRA. V9 region of the 18S rRNA gene: forward primer (1391F): GTACACACCGCCCGTC and reverse primer (EukB): TGATCCTTCTGCAGGTTCACCTAC.

Advances in next generation sequencing (NGS) technologies have resulted in its increased use in the field of microbial ecology. NGS is advantageous because the identification and relative abundance of a given microorganism in a microbial community can be obtained simultaneously and without the need for culturing the species, many of which remain uncultivable in the laboratory. Most of the commonly used sequencing technologies are PCR-dependent (usually referred to as sequencing by synthesis), requiring amplification steps prior to sequencing in order to increase the signal-to-noise ratio and to be compatible with the sequencing procedures. Examples of PCR-dependent NGS technologies are 454, PGM, Solid and Illumina. PacBio and Heliscope are PCR-independent sequencing technologies [8]. Metagenomics studies may be gene-specific or they can provide a picture of the overall community potential by sequencing all genes. The latter type of genome-wide metagenomics studies involve an amplification step using random primers prior to sequencing. Given that many NGS platforms are available, it is important to know how to choose a particular technology to solve a specific question. The main factors that guide this choice include the sequencing yield of the technology, sequencing coverage needed to make meaningful conclusions and to obtain adequate statistical power, error rate, read length, runtime, and cost of sequencing. It is hardly true that one technology fits the needs of all research questions that require NGS analyses, and the technologies, as well as their availability, keeps evolving. Here, we highlight specific uses, basic workflow, and pitfalls for the two commonly used platforms Illumina and PacBio. For more detailed information and for insights into additional platforms, we direct the reader to a review article by Quail et al. [9].

Currently, no technology has the ability to sequence an intact genome in its entirety. Instead, DNA fragments (e.g., amplicons generated using PCR or fragments generated using physical and chemical shearing) are obtained in the first step of sample preparation (also called library preparation), for both Illumina and PacBio platforms. Illumina library preparation for diversity analysis is a two-step process, which involves two independent amplifications. The first amplification (typically consisting of 25 cycles), called amplicon PCR, uses specific primers for the region of interest (e.g., 16S rRNA gene) that contain overhang adapters attached. The second amplification (typically consisting of 8 cycles) is needed to attach dual indices (unique indices to each sample; also called barcodes) and Illumina sequencing adaptors. The final amplicon should be between 300 and 500 bp for optimum results. For whole genome sequencing of pure culture (WGS) or a mixed community (metagenomics), DNA is first fragmented using physical (sonication) or chemical (enzymatic) methods followed by PCR using universal random primers to obtain genome-wide amplification prior to sequencing. Recently, PacBio sequencing is gaining momentum as choice technology because it allows a larger fragment size to be sequenced per sequence read. PacBio library preparation involves generating DNA fragments of up to 10 kb in size. In addition to this major difference in the length of sequence read, Illumina and PacBio also vary in the error rates, which are as low as 1 % for Illumina technologies and as high as ~15 % for PacBio sequencing.

Illumina MiSeq, HiSeq, and NextSeq are the most widely used NGS platforms between which the sequencing yield is the main distinguishing feature. Typical yields from MiSeq (up to 50 M paired end reads) adequately capture information required for most studies of microbial diversity (microbiome analysis) using 16S rRNA sequencing, and strategies have been developed that enable sequencing ~384 samples in a single run [10]. HiSeq and NextSeq technologies provide higher yields (between 400 M and 800 M paired end reads) and thus, are suitable for WGS or metagenomics analysis, as higher average sequence coverage (calculated as sequencing yield divided by genome size) can be obtained in either technology. Paired end reads (usually referred to as R1 and R2) represent the same DNA fragment sequenced from both directions (forward and reverse). Note that the DNA fragment may be larger than the length covered by the length of R1 + R2, in which case the middle portion of each fragment represents a gap in the sequence. Paired-end libraries provide information about the orientation of the fragments with respect to each other and are suitable for fragments of ~300–500 bp insert size (the length of DNA fragment between Illumina adapter sequences including R1 and R2 sequences), an option that is unavailable for PacBio technology. However, the strength of PacBio lies in generating each sequence of substantially larger length, thereby providing positional information and facilitating genome-wide assembly. The average yield of PacBio technology can be up to 300 Mb (per SMRT cell). As a result, PacBio also holds promise in conducting WGS and metagenomics analysis of small eukaryotes that have much larger genomes than many *Bacteria* and *Archaea*.

2   Materials

2.1   DNA Isolation

1. Phenol–chloroform–isoamyl alcohol (25:24:1).
2. Refrigerated centrifuge.
3. Microcentrifuge tubes.
4. Pipette (1000 µL) and pipette tips.
5. Chloroform–isoamyl alcohol (24:1).
6. Vortex.
7. 3 M sodium acetate (pH 5.2).
8. 100 % ethanol.
9. 70 % ethanol.
10. Freezer at −80 °C or at −20 °C.
11. Ultrapure water.
12. TE buffer: 1 mL of 1 M Tris–HCl, 200 µL of 0.5 M EDTA (pH 8.0) in 100 mL of ultrapure water and sterilize by autoclaving.

2.2   PCR

1. 10× PCR reaction buffer.
2. 25 mM $MgCl_2$.
3. dNTPs (10 mM each).
4. 10 µM forward primer.
5. 10 µM reverse primer.
6. Ultrapure water.
7. 5 U/µL Taq DNA polymerase.
8. Thermocycler.
9. 50× TAE buffer (pH 8.3): 242 g Tris base, 57.1 mL glacial acetic acid, and 100 mL 0.5 M EDTA (pH 8.0). Adjust to 1 L with dH_2O. Autoclave for 20 min. Store at room temperature. Dilute to obtain 1× TAE (electrophoresis buffer).
10. 1 % (w/v) agarose gel. Dissolve 1 g of agarose in 100 mL 1× TAE by heating in a microwave oven. Cool the agarose to 55–65 °C and add 3 µL of SYBR®Safe stain (*see* **Note 1**). Let it cool inside the tray with the comb.
11. Loading dye.
12. Electrophoresis apparatus.
13. UV transilluminator.

2.3   DGGE

1. Dcode™ Universal Mutation Detection System (Bio-Rad, EUA).
2. Gradient former.

3. Magnetic stirrer.

4. Peristaltic pump.

5. Glass plates: one large and one smaller.

6. Spacers (two).

7. Clamps (two).

8. Power supply.

9. Gelbond.

10. 96 % ethanol.

11. 70 % ethanol.

12. Plastic card (with the size of the gel).

13. Combs.

14. TEMED (Tetramethylethylenediamine).

15. 10 % ammonium persulfate solution in water.

16. 80 % denaturant solution, 8 % polyacrylamide: 200 mL 40 % acryl/bisacrylamide 37.5:1, 320 mL formamide, 10 mL 50× TAE buffer, 20 mL glycerol (only when silver staining is performed), 337.3 g urea. Carefully heat hand-warm, add stirrer and dissolve. Adjust to final volume of 1 L with dH_2O. Store in the dark and at room temperature.

17. 0 % denaturant solution, 8 % polyacrylamide: 200 mL 40 % acryl/bisacrylamide 37.5:1, 10 mL 50× TAE buffer, 20 mL glycerol (only when silver staining is performed). Adjust to the final volume of 1 L with dH_2O. Store in the dark at room temperature.

18. Steel boxes (two).

19. 1× Cairn's fixation solution. Prepare 8× fixation solution by adding 200 mL 96 % ethanol, 10 mL acetic acid, and 40 mL dH_2O. To prepare 1× Cairn's fixation solution add 50 mL of 8× solution and 350 mL dH_2O.

20. Silver staining solution. Add 0.4 g $AgNO_3$ to 200 mL of 1× Cairn's fixing solution.

21. Developer solution: a spatula tip of $NaBH_4$ (approx. 10 mg), 250 mL 1.5 % NaOH solution, and 750 µL formaldehyde.

22. Cairns' preservation solution: 250 mL 96 % ethanol, 100 mL glycerol, and 650 mL dH_2O.

23. Cellophane sheets.

24. Oven (60 °C).

2.4 T-RFLP

1. 10× PCR reaction buffer.

2. 25 mM $MgCl_2$.

3. dNTPs (2 mM of each nucleotide).

4. 10 µM forward primer.

5. 10 µM reverse primer.

6. Ultrapure water.

7. 10 U/µL Taq DNA polymerase.

8. Thermocycler.

9. Commercial kit for PCR products purification (e.g., Nucleo Spin Extract II, Macherey-Nagel; QIAquick PCR Purification Kit, QIAGEN Inc., Valencia, CA, or Exo-SAP protocol).

10. Restriction enzymes (*Alu* I, *Bst*UI, *Dde*I, *Sau*96I, *Msp*I, and *Hha*I).

11. 10× Restriction buffer.

12. Thermoblock.

13. Loading buffer.

14. Dye-labeled size standard.

2.5 Cloning for Sequencing

1. 10× PCR reaction buffer.

2. 25 mM $MgCl_2$.

3. dNTPs (10 mM each).

4. 10 µM forward primer.

5. 10 µM reverse primer.

6. Ultrapure water.

7. 5 U/µL Taq DNA polymerase.

8. Thermocycler.

9. 1 % (w/v) Agarose gel. Dissolve 1 g of agarose in 100 mL 1× TAE by heating in a microwave oven. Cool the agarose to 55–65 °C and add 3 µL of SYBR®Safe stain (*see* **Note 1**). Let it cool inside the tray with the comb.

10. Loading dye.

11. Electrophoresis apparatus.

12. UV transilluminator.

13. Commercial kit for PCR products purification (e.g., Nucleo Spin Extract II, Macherey-Nagel, Qiagen PCR purification kit).

14. Competent cells.

15. pGEM®-T Easy Vector System I (PROMEGA, catalog # A1360).

16. Laminar flow chamber.

17. LB medium: NaCl 10 g, tryptone 10 g, yeast extract 5 g. Add dH_2O until 1 L. Autoclave 121 °C, 20 min.

18. Ampicillin 100 mg/mL.

19. PCR 96-well plates and 96-deep well plates.

20. Gas-permeable adhesive seal and sterile adhesive PCR film.

21. Incubator.

22. TE buffer: 1 mL of 1 M Tris–HCl, 200 µL of 0.5 M EDTA (pH 8.0) in 100 mL of ultrapure water and sterilize by autoclaving.

23. Glycerol.

24. Freezer at −80 °C.

3   Methods

3.1   DNA Extraction Methods

The choice of the DNA extraction method is highly dependent on the type of biological sample such as soil, biofilms, plants, and blood for which there are specific commercial kits available (e.g., MO BIO Laboratories, MP Biomedicals, Qiagen, among others). DNA extraction, isolation, and purification is a critical step and the same methodology should be maintained for meaningful comparisons between samples within a study because different extraction methods yield differences in DNA quality, influencing downstream processing including PCR (e.g., cells in the biofilm and filamentous microbes are harder to lyse prior to DNA extraction, whereas a significant amount of water must be filtered to obtain sufficient cellular mass for an adequate DNA yield). Also, the samples, as well as the extracted DNA, should be preserved using appropriate techniques, as the preservation temperature, reagents and time can influence the microbial community composition (*see* **Note 2**). The first step in DNA extraction methods is the cell disruption. Cell lysis can be achieved mechanically (bead beating or equivalent), chemically (using detergents), by freeze and thaw, among others, as well as by combining two or more methods. After cell lysis, cell debris need to be separated from nucleic acids, usually by centrifugation. Together with nucleic acids, proteins will also remain in the supernatant and subsequent steps aim to isolate the nucleic acids. The classical method for nucleic acids isolation is extraction with phenol–chloroform–isoamyl alcohol. Most of the commercial kits do not use this method but it is still utilized in several laboratories. The principal steps of the phenol–chloroform–isoamyl alcohol method are the following:

1. To 500 µL of lysate/supernatant, add 500 µL of phenol–chloroform–isoamyl alcohol (25:24:1) and mix well.

2. Centrifuge at maximum speed for 5 min.

3. Carefully pipet the aqueous phase to a new tube, avoiding the interphase and the organic phase (phenol phase), if the phases are mixed, repeat the centrifugation step.

4. Add 500 µL of chloroform–isoamyl alcohol (24:1).

5. Vortex to homogenize.

6. Centrifuge at maximum speed for 3 min.

7. Carefully pipet the aqueous phase to a new tube.

8. Add 1/10 volume of 3 M sodium acetate (pH 5.2).

9. Add 2.5 volumes of 100 % ethanol, mix and spin down the sample.

10. Place at −80 °C for 30 min or at −20 °C, 2 h to overnight.

11. Centrifuge at 4 °C for 20 min to pellet DNA.

12. Carefully, pour off supernatant and wash the pellet with 70 % ethanol (cold).

13. Centrifuge at 4 °C for 3–5 min.

14. Pull off the ethanol with pipet tip and air dry the DNA pellet.

15. Resuspend the pellet in water or TE buffer.

After extraction, DNA quantity and quality should be evaluated. For DNA quantification, both spectrophotometric (e.g., NanoDrop) and fluorometric (e.g., Qubit) methods can be used. Loading the DNA in agarose gel electrophoresis is also a good way of evaluating DNA quality. Other systems, such as Bioanalyzer, can simultaneously provide information on the quantity and quality of the DNA isolated (*see* **Note 3**).

3.2 PCR Amplification

3.2.1 PCR Amplification General Procedure

Amplification of DNA by PCR requires a mixture of the 4 nucleotides (adenine, thymine, cytosine, and guanine), a forward and a reverse primer that complement the flanks of the DNA sequence to be amplified, Taq polymerase enzyme, which will incorporate the nucleotides and build the new DNA fragments, and magnesium ions that acts as a cofactor for the Taq polymerase enzyme. Several kits containing Taq polymerase are commercially available. The example in Table 1 is given for Taq DNA polymerase (recombinant) from Thermo (Ref.: EP0406).

Table 1

Reagents and respective volumes used in one polymerase chain reaction

10× PCR reaction buffer (without MgCl$_2$)	5 μL
25 mM MgCl$_2$	6 μL
dNTPs (10 mM each)	1 μL
10 μM forward primer	1 μL
10 μM reverse primer	1 μL
Water	34.75 μL
5 U/μL Taq DNA polymerase	0.25 μL
Total volume	49 μL

Usually, several samples are processed at the same time and a PCR premix is prepared by multiplying the volume per reaction by the total number of DNA samples to amplify. To each PCR mix reaction tube, 1–5 µL of DNA template (<500 ng/µL) is added. The extracted DNA can be used directly as a template or dilutions of original DNA can be used. When diluting DNA, possible existing contaminants (proteins, phenolic compounds, etc.) are diluted as well and therefore, the efficiency of DNA amplification may increase.

1. Prepare the PCR reactions as exemplified above and gently mix them before placing into the thermocycler.

2. Select the appropriate PCR program. The primer set U968 f (AAC GCG AAG AAC CTT AC)/L1401-r (CGG TGT GTA CAA GAC CC) [11] is an example that is often used to amplify bacterial DNA prior to DGGE and the PCR program suitable for this primer set includes initial denaturation at 95 °C for 2 min; 30–35 cycles of amplification (denaturation at 95 °C, 30 s, annealing at 56 °C, 40 s, extension at 72 °C, 1 min), and a final extension at 72 °C, 5 min, followed by a hold at 4 °C (*see* **Note 4** for common PCR troubleshooting). The amplification for DGGE should be performed using a GC-clamp (CGC CGG GGG CGC GCC CCG GGC GGG GCG GGG GCA) attached to the 5′ end of one of the primers [12].

3. Verify the size of the PCR products by comparison with appropriate size and mass standard using electrophoresis on a 1 % (w/v) agarose stained gel (up to 2 % gel is required for PCR products smaller than 500 bp). A SYBR® Safe stain or equivalent (such as ethidium bromide or GelRed™) can be applied, which intercalates into DNA and produces a fluorescent product when viewed under UV light (*see* **Note 1**).

4. Run gels at a constant voltage of 80–100 V for at least 15 min. Higher voltages lead to poorer separation, especially of fragments less than 500 bp in length and therefore smaller voltages are recommended for better separation of PCR products that are less than 500 bp in length. Nucleic acids are detected using a UV transilluminator.

3.3 DGGE

PCR amplified fragments that have the same nucleotide length can be separated in a denaturing gradient gel based on their nucleotide sequences. One of the primers used for amplifying the gene of interest should have a GC clamp attached in the 5′ end (a scheme of the DGGE technique is represented in Fig. 2).

3.3.1 DGGE Procedure

The procedure herein described is for 8 % (v/v) polyacrylamide gel running in a Dcode™ Universal Mutation Detection System.

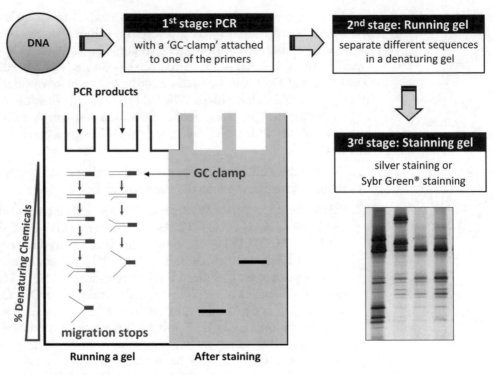

Fig. 2 Scheme of the DGGE technique

Assembling the DGGE
Plate Cassette

1. Clean one large and a smaller glass plate with soap, dry with paper and clean with 96 % ethanol.

2. Cut the gelbond to the size of the largest glass plate.

3. Add some drops of water to the surface of the large glass plate.

4. Place the gelbond hydrophobic side down on this glass plate.

5. Fix the gelbond, without removing the paper sheet, using a roller, then remove the paper sheet.

6. Dry the gelbond carefully using clean paper.

7. Clean a set of spacers with 96 % ethanol, and place them over the gelbond.

8. Place the smaller glass plate on top after ensuring that it is cleaned (this arrangement is called a sandwich).

9. Add the clamps to the sides of the sandwich, and place in the sandwich-holder. Verify the distance between the spacers by putting the plastic card (with the size of the gel) between the glass plates and the spacers.

10. Press the spacers down and fasten the screws on the clamps.

11. Place the sandwich on top of the rubber gasket and press down the handles.

Table 2

Denaturing gradient gel electrophoresis mixing table

Gradient percentage (%)	0 % Solution (mL)	80 % Solution (mL)	Volume (mL)	TEMED (µL)	10 % APS (µL)
0	9.0	0.0	13	13	50
30	8.1	4.9	13	13	50
40	6.5	6.5	13	13	50
50	4.9	8.1	13	13	50
60	3.3	9.8	13	13	50
70	1.6	11.4	13	13	50
80	0.0	13.0	13	13	50

Casting the Denaturing Gradient Gels

1. Prepare four gel solutions (HIGH, LOW, plug, and stacking gel) in 50 mL tubes, on ice, according to Table 2 (the APS solution is added just before pouring the gel). For bacteria and archaeal amplicons analysis, usually a 30–60 % gradient and a 30–50 % gradient are used, respectively. However, according to the result obtained gradients can be altered in order to get the best separation of the amplicons. Plug and stacking gels are prepared with 0 % solution.

2. Prepare 1 mL of plug gel solution (1.5 mL 0 % solution + 4.5 µL TEMED + 15 µL 10 % APS) and pour it at the bottom of the DGGE plate cassette (between glass plates) prior to making the main resolving gradient gels. It will prevent leakage from the bottom of the plates.

3. For casting gradient gels, a gradient former, a magnetic stirrer and a peristaltic pump is used.

4. Rinse the gradient maker and tubes with demi-water, switch on the pump at running speed (20 mL/min) and drain the system well.

5. Close the screw between the compartments of the gradient maker.

6. Dry the compartments with a tissue.

7. Add 10 % APS to the HIGH and LOW solutions and pour in the right and left compartment, respectively.

8. Start the stirrer and open the screw and immediately start the pump at 4 mL/min flow rate.

9. Place the needle between the glass plates and secure the tube.

10. Remove the needle when the gel is poured and switch off the pump.

11. Rinse the compartments with demi-water, switch on the pump and drain the system well.

12. Add the 10 % APS to the stacking gel.

13. Close the screw between the compartments and add the stacking gel to the right compartment.

14. Set the flow at 20 mL/min until the solution reaches the needle.

15. Change the flow to 1 mL/min and slowly increase to 4 mL/min.

16. When the stacking gel is poured, place the comb (previously cleaned with 70 % ethanol) carefully in the stacking gel. Avoid air-bubbles since these will appear as dents in all bands.

17. Leave the gel to polymerize for at least 1 h.

Running the Gel

1. Add 50× TAE buffer to the buffer tank and fill up until reaching the "fill" position.

2. Switch on the Dcode at least 90 min before electrophoresis starts, so that the buffer can heat up to 60 °C.

3. Remove the comb from the gel carefully.

4. Remove nonpolymerized gel fragments by rinsing the slots with demi-water and put the sandwich in the sandwich-holder (there should always be a sandwich at the other side to get a closed upper buffer compartment).

5. Switch off the Dcode, and take off the lid. Put the two sandwiches connected to sandwich holder inside the buffer tank.

6. Switch on the Dcode until the upper buffer compartment is filled with buffer.

7. Switch off the Dcode and take off the lid. Load your samples. About 1 μL of loading dye should be mixed with about 5 μL PCR product before loading the samples.

8. Put the lid and switch on the Dcode.

9. Switch on the power supply for 10 min at 200 V to guarantee that the samples enter into the gel. Then, reduce the voltage to 85 V and run for 16 h at 60 °C.

Staining the Gel

DGGE gels can be stained with SYBR®Safe or with silver (according to Sanguinetty et al. [13]), as following:

1. Place the gel in a stainless steel box.

2. Add 200 mL 1× Cairn's fixation solution and rock the box for 3 min, remove the solution and store for later application.

3. Add 200 mL silver staining solution to the box and rock for 10 min (use gloves).

4. Discard the solution (chemical waste).

5. Rinse the box with demi-water (chemical waste).

6. Add fresh demi-water and rock for 2 min.

7. Discard water (chemical waste).

8. Rinse the gel and gelbond with demi-water.

9. Place the gel in a stainless steel developer tray. Add a small part of developer solution and rock a little, discard this part and replace with the remaining developer solution.

10. Rock until the gel has developed well, which means that the bands are visible.

11. Discard the developer solution.

12. Add the previously used 200 mL 1× Cairns' solution to the box and rock for 5 min.

13. Discard the fixation solution.

14. Add demi-water and rock for 2 min.

15. Replace the demi-water with Cairns' preservation solution and rock for 7 min.

16. Place the gel on a glass plate (gelbond faced down).

17. Pre-wet cellophane foil in the preservation solution and put it over the gel (avoid air bubbles, press the edges of the gel to the glass plate).

18. Dry the gel overnight at 60 °C.

3.3.2 Comparative Analysis of DGGE Profiles

Most common methods for the comparative analysis of DGGE profiles include hierarchical cluster analysis, principal component analysis (PCA), multidimensional scaling (MDS) and canonical correspondence analysis (CCA). Gel Compar and Bionumerics (Applied Maths, St. Martens-Latem, Belgium) are examples of software available to perform the comparative analysis.

3.3.3 Interpretation of Ordination Analyses Using CCA

Ordination techniques including MDS, CCA, PCA and clustering algorithms aim at reducing the dimensionality of ecological data, which typically contain multiple environmental variables that may together determine microbial diversity. PCA is typically a biplot in which the two axes represent variance in species abundance, whereas CCA incorporates information from multiple regression analysis and is typically shown as a triplot in which the three axes represent variance in dispersion (or inertia). CCA) is the most commonly used technique that can provide information about how various environmental variables correlate with the species presence. Species and samples are shown as points and the most abundant species occupy the diagonal because of their largest contribution to inertia. Arrows from the origin indicate various "predictors" (environmental variables) of the distribution. The arrows at the right angle indicate independent effects; those at small angle

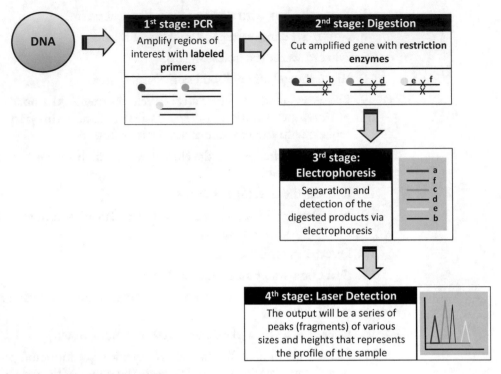

Fig. 3 Scheme of the T-RFLP technique

reflect positive correlation; whereas those pointing in the opposite directions represent negative correlation. Finally, the length of the arrow directly reflects the strength of the relationship.

3.4 T-RFLP

T-RFLP is a fingerprinting technique that is used to assess the composition and diversity of microbial communities (Fig. 3). At the end of a T-RFLP analysis, peaks of different sizes will compose the fragment profile of a given sample. Samples with different species compositions will present distinct T-RFLP profiles. Major steps of T-RFLP technique include DNA extraction and purification (*see* "DNA extraction methods"), PCR amplification followed by PCR product purification, digestion of PCR products with restriction enzymes, separation of fragments using electrophoresis, detection of terminal and fluorescently labeled restriction fragments, and comparison of profiles obtained from related samples.

3.4.1 T-RFLP Procedure

PCR Amplification

1. The choice of which gene to amplify is dependent on the objective of the experiment.is Typical diversity analysis experiments rely on 16S rRNA gene amplification. Table 3 lists some of the commonly used primers for DNA amplification prior to T-RFLP analysis. Primers can be labeled at the 5′ end with fluorescent dyes such as FAM (blue) or PET (red), among others.

Table 3

List of primers used for the analysis of microbial communities using T-RFLP

Primer name	Primer sequence (5′ - 3′)	Target
8F-Eub	AGAGTTTGATCCTGGCTCAG	Bacterial 16S rRNA gene
8fm-Eub	AGAGTTTGATCMTGGCTCAG	Bacterial 16S rRNA gene
49F-Eub	TNANACATGCAAGTCGRRCG	Bacterial 16S rRNA gene
334F-Eub	CCAGACTCCTACGGGAGGCAGGC	Bacterial 16S rRNA gene
341F-Eub	CCTACGGGAGGCAGCAG	Bacterial 16S rRNA gene
786F-Eub	GATTAGATACCCTGGTAG	Bacterial 16S rRNA gene
27F-Eub	GTTTGATCCTGGCTCAG	Bacterial 16S rRNA gene
519F-Univ	CAGCAGCCGCGGTAATAC	Universal 16S rRNA gene
536R-Eub	GWATTACCGCGGCKGCTG	Bacterial 16S rRNA gene
926R-Eub	CCGTCAATTCCTTTRAGTTT	Bacterial 16S rRNA gene
1113R-Eub	GGGTTGCGCTCGTTG	Bacterial 16S rRNA gene
1404R-Eub	GGGCGGWGTGTACAAGGC	Bacterial 16S rRNA gene
1511R-Eub	GYTACCTTGTTACGACTT	Bacterial 16S rRNA gene
1492R-Univ	CGGCTACCTTGTTACGAC	Bacterial 16S rRNA gene
1406R-Univ	ACGGGCGGTGTGTRC	Bacterial 16S rRNA gene
1511R-Eub	GYTACCTTGTTACGACTT	Bacterial 16S rRNA gene
1492R-Univ	CGGCTACCTTGTTACGAC	Universal 16S rRNA gene
1406R-Univ	ACGGGCGGTGTGTRC	Universal 16S rRNA gene

Primer specificity can be checked using informatics tools such as Primer Prevalence (http://mica.ibest.uidaho.edu/primer.php) or Probe Match (https://rdp.cme.msu.edu/probematch/search.jsp).

2. The master mix composition for PCR amplification is shown in Table 4. The volumes presented are for one reaction and should be multiplied by the number of reactions to perform. It is also customary to scale down the reaction to a total volume of 25 μL or 12.5 μL.

3. The PCR program will be dependent on the specific polymerase used and the primer sequences, which determine the melting temperature (T_m) of the primer. The following PCR program can be used for the primers U27f and U1492r (*see* Table 3): initial denaturation at 95 °C for 5 min; 25–30 cycles of amplification (95 °C, 40 s; 56 °C, 40 s; 72 °C, 1.5 min), and a final extension at 72 °C, 10 min, followed by a hold at 4 °C. This program can be modified depending upon the

Table 4

Master mix composition for PCR amplification

5 μL	10× PCR reaction buffer
1.5 μL	50 mM MgCl$_2$
5 μL	dNTPs (2 mM of each nucleotide)
1 μL	10 μM forward primer
1 μL	10 μM reverse primer
0.5 μL	10 U/μL Taq polymerase
31 μL	H$_2$O
5 μL	Target DNA
50 μL	Total volume

polymerase enzyme being used. Most polymerases (e.g., Takara Extaq) come with a recommended program that may vary from the one mentioned above. The size of PCR products should be checked in agarose gel electrophoresis (*see* "PCR amplification" section) using appropriate DNA ladders.

3.4.2   PCR Product Purification

PCR products can be purified using commercially available kits following the manufacturers' instructions.

3.4.3   Restriction Enzyme Digestion

Most common restriction enzymes used to study microbial communities are *Alu*I, *Bst*UI, *Dde*I, *Sau*96I, *Msp*I, and *Hha*I. Digestion can be checked in silico using the tools available at http://mica. ibest.uidaho.edu/   or   http://nc2.neb.com/NEBcutter2/ (*see* **Note 5**).

1. Mix the following components in a microtube.

200–400 ng	Purified PCR product
1,5 μL	10× restriction buffer
10 U	Restriction enzyme
15 μL	Total volume (with H$_2$O)

2. Incubate the tubes at restriction enzyme's optimal temperature (follow manufacturer's instructions).

3. Heat for 20 min at 75 °C to inactivate the enzymes.

3.4.4   Electrophoresis

An aliquot of digested PCR product (2 μL) is mixed with loading buffer (2.5 μL) and dye-labeled size standard (0.5 μL), and denatured by incubation at 94 °C for 2–4 min. Samples are ready to be loaded in the electrophoresis system using the manufacturer's instructions.

3.4.5 Data Analysis

When forward and reverse primers are fluorescently labeled, two profiles per sample will be generated and if two restriction enzymes are used, each sample will have a total of four profiles. There is software available to analyze the peak patterns generated in T-RFLP assays (e.g., GeneMapper, Applied Biosystems). Additional data analysis such as cluster analysis, PCA, MDS, CCA, or redundancy analysis (RDA) can be performed to understand similarities between samples and the correlations with environmental factors.

3.4.6 Microbial Community Identification

T-RFLP results can be combined with cloning and sequencing in order to identify the predominant microorganisms. Taxonomic identity of microorganisms can also be retrieved from T-RFLP data using informatics tools such as TAP-TRFLP, MiCA, T-RFLP Phylogenetic Assignment Tool (PAT), TReFID, TRAMPR, ARB-software integrated tool (TRF-CUT), TRiFLe, T-RFPred.

3.5 Cloning for Sequencing

Fingerprinting methods allow comparing the diversity of mixed microbial communities but give no information on the taxonomical identity of microorganisms. Cloning and sequencing methodologies can be coupled to fingerprinting methods in order to link microbial diversity and identification. Fig. 4 represents the principal steps of the cloning procedure. PCR amplified genes are incorporated in a cloning vector (e.g., plasmid) that is then inserted in a competent cell (usually *Escherichia coli*). Because the plasmid has a gene conferring resistance to antibiotics (usually ampicillin), *E. coli*

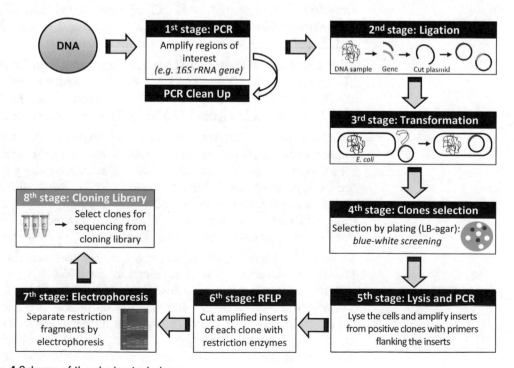

Fig. 4 Scheme of the cloning technique

cells containing the plasmid will be able to grow on antibiotic-containing medium but the ones without the plasmid will not grow on the selection medium. Each plasmid will incorporate one PCR fragment only. Therefore, each *E. coli* colony, growing in the selection medium, contains a gene from a single microbe that is present in the original microbial community. The genes isolated by cloning can be amplified and sequenced subsequently. Sequences may be compared to those present in databases in order to obtain microbial identification.

3.5.1 PCR Amplification and Cloning

1. 16S rRNA genes are amplified with specific primers (e.g., U27f and U1492r) using the following PCR program: initial denaturation at 95 °C for 2 min; 30 cycles of amplification (95 °C, 30 s; 52 °C, 40 s; 72 °C, 1.5 min), and a final extension at 72 °C, 5 min, followed by a hold at 4 °C until PCR products can be visualized in a 1 % agarose gel by mixing 5 µL of PCR product with 1 µL of loading dye.

2. PCR products are purified prior to inserting in the cloning vector, using commercially available kits or alternatively, by following the traditional ethanol precipitation protocol.

3. Purified PCR products are cloned into the pGEM-T easy vector kit according to the manufacturers' instructions.

3.5.2 Screening for Transformants by Colony PCR

1. Individual white colonies are inoculated in each well of a 96 deep well plate, containing LB medium supplemented with ampicillin (0.1 mg/mL), at 37 °C, 250 rpm for at least 17 h. The plate should be covered with a gas-permeable adhesive seal for growing *E. coli* cells.

2. After growth, each clone (10 µL) is transferred to TE buffer (90 µL) and incubated 10 min at 95 °C. Cell lysates may be stored at −20 °C and clones may be preserved in glycerol at −80 °C (add 15 µL of sterile glycerol (50 %) to 35 µL of each clone).

3. To check the correct size of the insert sequence, plasmids sequences are amplified from 1 µL lysed clones using the primer set PG1f/PG2r provided in the pGEM-T vector kit and using the PCR program: initial denaturation at 95 °C for 2 min; 30 cycles of amplification (95 °C, 30 s; 55 °C, 40 s; 72 °C, 1.5 min), and a final extension at 72 °C, 5 min, followed by a hold at 4 °C. PCR products can be visualized in a 1 % agarose gel.

4. In order to identify which insert sequence corresponds to DGGE bands of interest, amplification for DGGE analysis should be performed, as described in Subheading 3.2. Sequences separated by cloning and the sequences present in the initial DGGE profile can be loaded in different lanes of the same DGGE gel in order to get the correspondence and decide which insert should be sequenced to obtain the identification.

3.6 Next Generation Sequencing

Metagenomic studies are commonly performed by analyzing the prokaryotic 16S ribosomal RNA gene (16S rRNA), which is approximately 1500 bp long and contains nine variable regions interspersed between conserved regions. Variable regions of 16S rRNA are frequently used in phylogenetic classifications such as genus or species in diverse microbial populations. Which 16S rRNA region to sequence is an area of debate, and your region of interest might vary depending on things such as experimental objectives, design, and sample type. A protocol describing a method for preparing samples for sequencing the variable V3 and V4 regions of the 16S rRNA gene using the Illumina technology can be obtained in the Illumina website [14]. The protocol can also be applied to sequence alternative regions of the 16S rRNA gene and for other targeted amplicon sequences of interest.

A detailed overview of amplicon sequencing using PacBio technology is available at the PacBio website [15, 16].

4 Notes

1. GelRed™ stain (Biotium #41003) is the preferred alternative to SYBR®Safe because of its sensitivity and nonmutagenic properties compared to other commonly used alternatives such as ethidium bromide.

2. Preservation of samples for optimal DNA yield and quality. One of the challenges in the analysis of microbial diversity using PCR-dependent methods is that it is not always possible to directly isolate DNA from the collected samples, requiring one to preserve the samples prior to DNA extraction and PCR. Various preservation methods and reagents are available and reviewed elsewhere [17–19]. Overall, there are three approaches to preserving samples.

   (a) Freezing: samples can be frozen at −80 °C or in dry ice indefinitely prior to DNA extraction. One caveat is that this approach requires expensive instrumentation, which is not always available. Also, freeze-thawing before DNA extraction can lead to DNA degradation.

   (b) Drying: samples are dried as fast as possible (freeze-drying is recommended) and stored at room temperature prior to DNA extraction.

   (c) Buffering: samples are mixed with a DNA stabilization reagent such as sterile PBS or glycerol and stored at room temperature for a few weeks to a month. Buffering reagents are also commercially available as parts of preservation kits (e.g., Norgen Biotek Corp. # 27650).

3. Nucleic acid quantification. Various instruments are available for measuring DNA concentration prior to PCR and sequencing. The commonly used technologies are described below.

   (a) Gel electrophoresis: this is the least instrument-intensive and the most time consuming method among all. The nucleic acids are loaded on agarose gel along with markers (usually genomic DNA of lambda phage) at specific known concentrations. Although it is difficult to measure an exact concentration, gel electrophoresis can be used to reliably establish the range of concentration as well as verify the quality. Degraded nucleic acids appear as smears instead of intact bands on the gel.

   (b) NanoDrop technology (Thermo Scientific): this is the most commonly used method for quick DNA quantification. It uses spectrophotometric principles. Advantages include relatively competitive pricing and requirements of small amounts (1 μL) of DNA without further processing, and also retrieves 260/280 and 260/230 ratios which give information on DNA/RNA purity (260/280 ratios lower than ~1.8 for DNA and ~2 for RNA may indicate contamination by phenol, proteins or other contaminants that absorb near 280 nm; 260/280 ratios lower than 2.0–2.2 may indicate the presence of contaminants that absorb near 230 nm). A major disadvantage includes typically overestimates of nucleic acids that are inconsistent with other methods.

   (c) Qubit technology (Thermo Scientific): currently, Qubit is the most accepted method for routine nucleic acid quantification. It is a fluorometric method. High sensitivity as well as broad range kits are available for both DNA, RNA and protein quantification. Small volumes can be used and the method is quick; however, the kits add significantly to the cost compared to the Nanodrop technology.

   (d) Bioanalyzer technology (Agilent): this is the method of choice for measuring DNA and RNA concentration and quality prior to next generation sequencing. Bioanalyzer is the most expensive, sensitive, and reliable method of quantification available today. See manufacturer's manual for more details [20].

4. Troubleshooting PCR. A polymerase chain reaction that yields no product impedes the rest of the analysis and can be frustrating. Factors frequently responsible for a failed PCR are described below.

   (a) Melting temperature of the primer (T_m): it is necessary that the T_m of the forward and reverse primers (temperature at which at least 50% of the primer is single stranded) is comparable. Most reactions will be successful at ~56 °C. Temperature much lower than this optimum may cause misspecific primer binding leading to false positive reactions; higher temperatures may increase denaturation and decreased efficiency. The optimal T_m depends on the length and sequence of the primers and can also be determined by doing PCR across a gradient of annealing temperatures.

(b) Extension timing: the length of the expected PCR product guides the extension time in a PCR. As a rule, for every 500 bp, an additional 30 s is necessary at the extension step during PCR cycles.

(c) Inhibitors: DNA extracted from environmental samples may contain inhibitors that can interfere with template–primer interaction during PCR. Dilution of DNA template often helps; however, more difficult samples can be amplified using buffers such as DMSO (at 1% final concentration) and bovine serum albumin (1.25 μL of the commercially available BSA in 12.5 μL total reaction volume).

5. It is important to select the best restriction enzymes taking into account the primers used in the amplification. Enzymes which result in the highest number of peaks should be selected. In order to have a better identification of the members of a given community, as well as a better distinction between different communities, it is advised to perform restriction profiles with more than one enzyme.

References

1. VerBerkmoes NC, Denef VJ, Hettich RL, Banfield JF (2009) Systems biology: functional analysis of natural microbial consortia using community proteomics. Nat Rev Microbiol 7(3):196–205

2. Hugerth LW, Muller EE, Hu YO, Lebrun LA, Roume H, Lundin D, Wilmes P, Andersson AF (2014) Systematic design of 18S rRNA gene primers for determining eukaryotic diversity in microbial consortia. PLoS One 9(4):e95567

3. Stock M, Lampert KP, Moller D, Schlupp I, Schartl M (2010) Monophyletic origin of multiple clonal lineages in an asexual fish (*Poecilia formosa*). Mol Ecol 19(23):5204–5215

4. Costas BA, McManus G, Doherty M, Katz LA (2007) Use of species-specific primers and PCR to measure the distributions of planktonic ciliates in coastal waters. Limnol Oceanogr Methods 5(6):163–173

5. Martin KJ, Rygiewicz PT (2005) Fungal-specific PCR primers developed for analysis of the ITS region of environmental DNA extracts. BMC Microbiol 5:28

6. Borneman J, Hartin RJ (2000) PCR primers that amplify fungal rRNA genes from environmental samples. Appl Environ Microbiol 66(10):4356–4360

7. Massana R, del Campo J, Sieracki ME, Audic S, Logares R (2014) Exploring the uncultured microeukaryote majority in the oceans: reevaluation of ribogroups within stramenopiles. ISME J 8(4):854–866

8. Buermans HP, den Dunnen JT (2014) Next generation sequencing technology: advances and applications. Biochim Biophys Acta 1842(10):1932–1941

9. Quail MA, Smith M, Coupland P, Otto TD, Harris SR, Connor TR, Bertoni A, Swerdlow HP, Gu Y (2012) A tale of three next generation sequencing platforms: comparison of Ion Torrent, Pacific Biosciences and Illumina MiSeq sequencers. BMC Genomics 13:341

10. Kozich JJ, Westcott SL, Baxter NT, Highlander SK, Schloss PD (2013) Development of a dual-index sequencing strategy and curation pipeline for analyzing amplicon sequence data on the MiSeq Illumina sequencing platform. Appl Environ Microbiol 79(17):5112–5120

11. Nübel U, Engelen B, Felske A, Snaidr J, Wieshuber A, Amann RI, Ludwig W, Backhaus H (1996) Sequence heterogeneities of genes encoding 16S rRNAs in *Paenibacillus polymyxa* detected by temperature gradient gel electrophoresis. J Bacteriol 178(19):5636–5643

12. Muyzer G, de Waal EC, Uitterlinden AG (1993) Profiling of complex microbial populations by denaturing gradient gel electrophoresis analysis of polymerase chain reaction-amplified genes coding for 16S rRNA. Appl Environ Microbiol 59(3):695–700

13. Sanguinetti CJ, Dias Neto E, Simpson AJ (1994) Rapid silver staining and recovery of PCR products separated on polyacrylamide gels. Biotechniques 17(5):914–921

14. Illumina (2016) 16S metagenomic sequencing library preparation. http://support.illumina.com/downloads/16s_metagenomic_sequencing_library_preparation.html. Accessed 30 Jul 2016

15. Pacific Biosciences (2016) PacBio® Procedure & Checklist - Amplicon Template Preparation and Sequencing. http://www.pacb.com/wp-content/uploads/Procedure-Checklist-Amplicon-Template-Preparation-and-Sequencing.pdf. Accessed 30 Jul 2016

16. Pacific Biosciences (2016) PacBio® Procedure & Checklist - Target Sequence Capture Using SeqCap® EZ Libraries with PacBio® Barcoded Adapters. http://www.pacb.com/wp-content/uploads/Procedure-Checklist-Target-Sequence-Capture-Roche-NimbleGen-SeqCapEZ-Library-PacBioBarcodedAdapters.pdf. Accessed 30 Jul 2016

17. Lauber CL, Zhou N, Gordon JI, Knight R, Fierer N (2010) Effect of storage conditions on the assessment of bacterial community structure in soil and human-associated samples. FEMS Microbiol Lett 307(1):80–86

18. Kerckhof FM, Courtens EN, Geirnaert A, Hoefman S, Ho A, Vilchez-Vargas R, Pieper DH, Jauregui R, Vlaeminck SE, Van de Wiele T, Vandamme P, Heylen K, Boon N (2014) Optimized cryopreservation of mixed microbial communities for conserved functionality and diversity. PLoS One 9(6):e99517

19. Saito MA, Bulygin VV, Moran DM, Taylor C, Scholin C (2011) Examination of microbial proteome preservation techniques applicable to autonomous environmental sample collection. Front Microbiol 2:215

20. Agilent Technologies (2005) Agilent 2100 bioanalyzer 2100 expert user's guide. https://www.agilent.com/cs/library/usermanuals/Public/G2946-90004_Vespucci_UG_eBook_(NoSecPack).pdf. Accessed 30 Jul 2016

Chapter 17

PCR in Metagenomics

Tina Kollannoor Johny and Sarita Ganapathy Bhat

Abstract

Metagenomics approach involves direct genetic analysis of environmental samples, evading the tedious culturing process. Polymerase chain reaction is one invaluable tool used for such analyses. Here, we describe one protocol for metagenomic DNA isolation that gives inhibitor-free DNA suitable for PCR and other genetic manipulations. Subsequently, the chapter describes the use of PCR as an indicator of quality of DNA and to amplify a marker of phylogeny. Further, the application of PCR for detection of specific genes and screening of metagenomic libraries is outlined.

Key words Metagenomics, 16S rDNA, Phylogeny, ARDRA profiling, Metagenomic library, Colony PCR

1 Introduction

Metagenomics refers to the direct genetic analysis of members of a microbial community. The approach is based on direct genetic analysis of the environmental samples, circumventing the tedious culturing of organisms and their individual study. Metagenomic analysis may be based on cloning and/or expression of DNA (functional metagenomics) or direct sequence analysis (sequence based metagenomics) [1]. Irrespective of the strategy involved, it is almost impossible to conceive metagenomics studies evading the invaluable technique of Polymerase Chain Reaction (PCR). PCR can be used at all stages of metagenomic analyses, as an indicator of purity of metagenomic DNA preparation and to answer the questions beginning from "Who is there?" "What are they doing?" and "Who is doing what?"

Metagenomic DNA extraction is the most challenging step in metagenomics studies because it is difficult to obtain good quality DNA of sufficient quantity [2]. The quality of DNA is critical for subsequent analyses such as cloning or PCR. We compared several existing metagenomic DNA extraction protocols [3] and developed

Lucília Domingues (ed.), *PCR: Methods and Protocols*, Methods in Molecular Biology, vol. 1620, DOI 10.1007/978-1-4939-7060-5_17, © Springer Science+Business Media LLC 2017

a new protocol that gives good success with metagenomic DNA isolation from fish gut and other biological samples. It provides DNA of good purity and is routinely used in our lab for cloning and PCR. The amplification of phylogenetic marker 16S rDNA can be used to assess the quality of metagenomic DNA [3–5] and to infer phylogeny [6, 7]. As community DNA is the subject matter of interest here, it is important to separate the mixture of PCR products, as each might represent a different bacterial taxon. This is done by cloning into a TA cloning vector and each clone is subjected to plasmid isolation and analysis of the insert. PCR also finds application in gene prospecting. It can be used for direct amplification of specific genes based on the sequences available in the public repositories [8] or metagenomic libraries can be screened using colony PCR [9], thereby evading the tedious functional screening process.

2   Materials

Perform all the experiments on ice, unless otherwise stated. All centrifugation steps should be done at 4 °C. Gloves should be worn at all times to prevent contamination with DNases. Use chemicals and pipettes dedicated to molecular biology for DNA isolation and PCR. Use sterile nuclease-free water for the preparation of all reagents. In case of enzymes and dNTPs, avoid multiple freeze–thaw cycles to prevent denaturation.

2.1 Isolation of Metagenomic DNA

1. Environmental sample: fish gut (in this case) or soil or any other biological material, collected and stored under sterile conditions.

2. Extraction buffer: 100 mM Tris–HCl (pH 8.2), 100 mM EDTA (pH 8), 1.5 M NaCl. Prepare 50 mL buffer and sterilize by autoclaving.

3. Cheese cloth.

4. Whatman filter paper (Grade 1 and 5). Diameter of the filter circles must be based on the size of the filtration apparatus. We use 47 mm filters.

5. 1× TE buffer (pH 8): 10 mM Tris–HCl, 1 mM EDTA.Na$_2$. Sterilize by autoclaving or filtration.

6. Cooling centrifuge: all centrifugations are to be carried out at 4 °C.

7. 50 mg/mL lysozyme solution.

8. 10 mg/mL RNase solution.

9. 10% w/v SDS solution.

10. 10 mg/mL proteinase K.

11. 5% w/v CTAB.

12. InhibitEX buffer (Qiagen).

13. Phenol–chloroform–isoamyl alcohol, 25:24:1 v/v/v.

14. Chloroform–isoamyl alcohol, 24:1 v/v.

15. 7.5 M potassium acetate solution.

16. Isopropanol.

17. 70% ethanol.

2.2   Agarose Gel Electrophoresis

1. Molecular grade agarose.

2. 10× TAE buffer (100 mL): 24.2 g Tris base, 5.7 mL glacial acetic acid, 10 mL of 0.5 M EDTA (pH 8), diluted to 100 mL using double distilled water. Sterilize the 10× concentrate by autoclaving and dilute to 1× concentration with sterile water before use.

3. 6× agarose gel loading dye: 0.25% w/v bromophenol blue, 0.25% w/v xylene cyanol FF, 40% w/v sucrose. Store at 4 °C.

4. DNA marker: Lambda DNA/*Eco*RI plus *Hind*III Marker. Other DNA markers may also be used.

5. Ethidium bromide or SYBR safe. Ethidium bromide is prepared at a concentration of 10 mg/mL in sterile water and stored in dark air tight container. As it is a potent mutagen, we use the less hazardous alternative, SYBR Safe DNA gel stain. It is available as a 10,000× concentrate in DMSO. Dilute it to 1× concentration using 1× TAE buffer.

2.3   16S rDNA PCR

1. Metagenomic DNA of good purity, obtained using the metagenomic DNA isolation protocol given below or commercial kits.

2. Universal primers for 16S rDNA: 5′AGAGTTTGAT CCTGGCTCAG 3′ and 5′ ACGGCTACCTTGTTACGACTT 3 [10].

3. Sterile nuclease-free water.

4. 10× PCR buffer (Sigma-Aldrich).

5. 25 mM magnesium chloride solution.

6. dNTP mix (10 mM each).

7. Taq DNA polymerase (5 U/μL).

2.4   ARDRA Profiling and Determination of the 16S rDNA Clones Identity

1. pGEM-T Vector system (Promega). Other TA cloning systems may also be used.

2. *E coli* JM109 cells.

3. Transform Aid Bacterial Transformation Kit (Thermo Fisher Scientific).

4. LB agar (Composition/L): 10 g casein enzymic hydrolysate, 5 g yeast extract, 10 g sodium chloride, 20 g agar, Final pH (at 25 °C) 7.5 ± 0.2.

5. 50 mg/mL ampicillin.

6. 20 mg/mL X-gal in DMSO.

7. 20 mg/mL IPTG.

8. AXI plate: Add 100 μL of ampicillin to LB agar before pouring onto petri dishes and spread 40 μL each of X-gal and IPTG per plate.

9. QIAprep Spin Miniprep Kit (Qiagen).

10. Restriction enzymes: *Hpa*II and *Hha*I (10 U/μL) and their buffers.

11. Online bioinformatics tools: BLAST, MEGA 6.0 software, RDP Naïve Bayesian rRNA classifier.

2.5 Detection of Specific Genes in the Metagenome

1. EmeraldAmp GT PCR Master Mix (2× Premix).

2. Metagenomic DNA (Template).

3. Forward and reverse primers (10 μM each). For instance, our lab uses the primer pair specific for asparginase [11] and several custom-made primers.

Asn-F: (5′-ATCGGGATCCATGGAAAGAAAACACATTTAC-3′)

Asn-R: (5′ATCGGAATTCTTTTAGTGAGTTAACTCACCC-3′)

4. Reagents for TA cloning as described in Subheading 2.4.

5. Silica bead DNA gel extraction kit (Thermo Fisher Scientific).

2.6 Vector-Specific PCR

1. M13 forward primers 5′-d(GTTTTCCCAGTCACGAC)-3′ or 5′-d(CGCCAGGGTTTTCCCAGTCACGAC)-3′.

2. M13 reverse primers: 5′-d(CAGGAAACAGCTATGAC)-3 or 5′-d(CAGGAAACAGCTATGAC)-3.

3. SP6 primer: 5′-d(TATTTAGGTGACACTATAG)-3′.

4. T7 primer: 5′-d(TAATACGACTCACTATAGGG)-3′.

3 Methods

3.1 Isolation of Metagenomic DNA

1. Suspend the guts in 10 mL of extraction buffer (*see* **Notes 1–4**) and vortex for 1 min.

2. Centrifuge the suspension at 1600 × g for 4 min and collect the supernatant into a new tube.

3. Add 20 mL of extraction buffer to the pellet of gut contents, vortex again for 1 min, and centrifuge the suspension at 900 × g for 3 min.

4. Collect the supernatant and add it to the previously collected supernatant.

5. Add the remaining 20 mL of extraction buffer to the pellet, vortex for 1 min, and centrifuge the suspension at $900 \times g$ for 3 min.

6. Collect the supernatant and add it to the previously collected supernatant.

7. Filter the supernatant through 3–4 layers of cheese cloth.

8. Collect the filtrate, filter it through Whatman filter (Grade 1) under vacuum, and collect the new filtrate

9. Filter the new filtrate through 1.2-μm filter (Grade 5) under vacuum and collect the final filtrate.

10. Centrifuge the final filtrate at $15,000 \times g$ for 20 min and discard the supernatant.

11. Resuspend the pellet in 1 mL of 1× TE buffer.

12. Aliquot 500 μL into a 2 mL microtube and add 50 μL lysozyme and 10 μL of RNase solution.

13. Incubate the mixture at 37 °C for 1 h.

14. Add 250 μL SDS solution and 100 μL of proteinase K, and incubate at 55 °C for 1 h.

15. Add 100 μL of CTAB solution and incubate at 55 °C for 10 min.

16. Add equal volume of InhibitEX buffer and incubate at 55 °C for 10 min.

17. Centrifuge the suspension and collect the supernatant.

18. Extract the aqueous phase with equal volume of phenol–chloroform–isoamyl alcohol by centrifugation at $12,000 \times g$ for 10 min.

19. Collect the aqueous phase and extract with equal volume of chloroform–isoamyl alcohol.

20. Collect the aqueous phase, add one-tenth volume of 7.5 M potassium acetate and equal volume of isopropanol, and incubate at room temperature for 10 min.

21. Centrifuge the mixture at $12,000 \times g$ for 20 min and pour off the supernatant without disturbing the pellet.

22. Add 100 μL of 70% ethanol and centrifuge at $12,000 \times g$ for 2 min.

23. Remove the liquid and repeat the process.

24. Air-dry the pellet and resuspend it in 40 μL of 1× TE buffer.

25. Quantify the DNA using a spectrophotometer or a nano spectrophotometer (*see* **Notes 5** and **6**). Take the readings at 260, 280, and 230 nm. These readings are indicative of DNA quality (*see* **Note 7**).

3.2 Agarose Gel Electrophoresis

1. Prepare a mini gel of 0.8% agarose in 1× TAE buffer. Heat to dissolve the contents and allow it to cool for some time. Pour the mixture into the casting tray with the comb kept in position. Once solidified, transfer the gel to the gel tank containing 1× TAE buffer, sufficient to submerge the gel by a few millimeters.

2. Aliquot 5 µL of the DNA, mix with 1 µL of 6× gel loading dye, and load the mixture into the wells.

3. Run the gel at a constant voltage of 60 V for 1 h.

4. Stain the gel in ethidium bromide or SYBR safe.

5. Photograph the gel using any gel documentation system (Fig. 1).

3.3 16S rDNA PCR

1. Prepare a master mix consisting of 2 µL of 10× PCR buffer, 2 µL of 10 mMdNTP mix (200 µM), 1 µL of forward and reverse primers (0.02 mM each), 0.4 µL of 25 mMMgCl$_2$ (2 mM), and 0.2 µL of 5 U/µL Taq polymerase. Make up the volume to 19 µL using sterile nuclease-free water (*see* **Notes 8** and **9**).

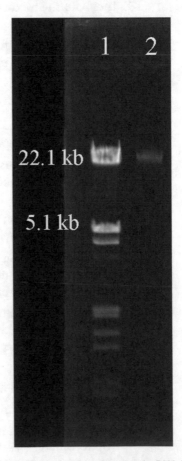

Fig. 1 Agarose gel electrophoresis of metagenomic DNA isolated from fish gut. *Lane 1*: Lambda DNA/*Eco*RI plus *Hind*III Marker, *Lane 2*: Metagenomic DNA

2. Aliquot the master mix to required number of clearly labeled microtubes.

3. To all the tubes, add 1 µL of metagenomic DNA, whose concentration is adjusted to 50 ng.

4. Mix the contents of each tube and spin the tubes briefly to bring all the contents to the bottom of the tube (*see* **Note 10**).

5. Place the tubes in a thermal cycler and set up the following cycling program:

94 °C, 5 min (Initial denaturation),

Followed by 34 cycles of

94 °C, 0.5 min.

56 °C, 0.5 min.

72 °C, 1 min.
72 °C, 10 min (Final extension).

6. Analyze the PCR products on 1.2% agarose gel as described in Subheading 3.2 (*see* **Note 8**). (Fig. 2)

3.4 ARDRA Profiling and Determination of the 16S rDNA Clones Identity

1. Ligate the PCR products into the pGEM-T Vector (Fig. 3). Set up the following ligation mix:

2× Rapid Ligation buffer	5 µL
pGEM-T Vector (50 ng)	1 µL
PCR product (75 ng/µL)	1 µL
T4 DNA Ligase (3 Weiss Units/µL)	1 µL
Nuclease-free water	2 µL

2. Mix the contents properly and incubate overnight at 4 °C (*see* **Notes 11–13**).

3. Transform 2.5 µL of ligation mix into 50 µL of competent *E coli* JM109 cells prepared using the Transform Aid Bacterial Transformation Kit according to the manufacturer's instructions (*see* **Notes 14** and **15**).

4. Plate the cells on LB agar supplemented with ampicillin, X-gal, and IPTG (AXI plate), and incubate overnight at 37 °C.

5. Select the white clones and patch them on LB agar plates supplemented with ampicillin. This constitutes the phylogenetic library of the sample.

6. Inoculate the clones into LB broth supplemented with the same final concentration of ampicillin and incubate for 12–16 h with shaking at 250 rpm.

7. Isolate plasmids from the phylogenetic clones using QIAprep Spin Miniprep Kit, following the manufacturer's instructions (*see* **Note 16**).

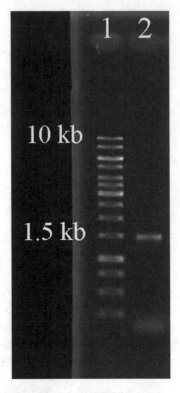

Fig. 2 Agarose gel electrophoresis of 16S PCR amplicons. *Lane 1*: GeneRuler 1-kb DNA Ladder, *Lane 2*: 16S PCR amplicon

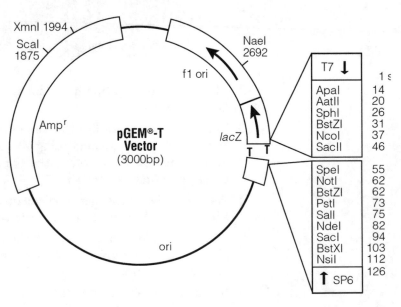

Fig. 3 pGEM-T Vector map

8. Re-amplify the 16S rDNA from the plasmids by using the same 16S PCR (*see* Subheading 3.3.). However, use 1 μL of plasmid DNA (50 ng) as the template. Analyze the amplicons on 1% agarose gel as described above.

9. Digest the amplicons, individually with 5 U of restriction endonuclease *Hpa*II. Set up the following reaction:

PCR reaction mixture	10 μL (0.1–0.5 μg of DNA)
Nuclease-free water	17.5 μL
10× Buffer Tango	2 μL
*Hpa*II	0.5 μL

Mix gently and incubate for 4 h at 37 °C.

10. Analyze the digest on 2% agarose gel and image the gel using any gel documentation system.

11. Group the ARDRA patterns (*see* **Note 17**) and digest clones with similar patterns using *Hha*I. Set up the reaction as explained in **step 9**. Use 0.5 μL of *Hha*I instead of *Hpa*II (Fig. 4).

12. Sequence the clones showing diverse patterns by Sanger's dideoxy method.

13. Determine the identity of the sequence using nucleotide BLAST of the NCBI database.

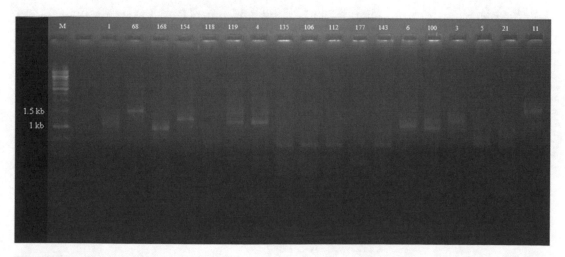

Fig. 4 Agarose gel electrophoresis of 16S rDNA plasmid library restriction with *Hpa*II. *Lane 1*: GeneRuler 1-kb DNA Ladder, Other lanes identified with *clone numbers*

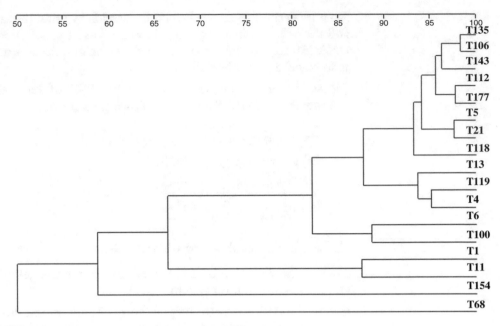

Fig. 5 Phylogenetic tree constructed based on ARDRA pattern

14. Compile and align the sequences using the Clustal W program of the MEGA software. Construct a phylogenetic tree using the Neighbor Joining method of the MEGA software (Fig. 5).

15. Assign taxonomic hierarchy to the sequence using the RDP Naïve Bayesian rRNA classifier. The software groups sequences into different phyla and thus helps to establish the phylogenetic diversity of the metagenome sample (*see* **Note 9**).

3.5 Detection of Specific Genes in the Metagenome

1. Aliquot 25 μL of EmeraldAmp GT PCR Master Mix (*see* **Notes 18** and **19**) into a new labeled microtube.

2. Add 1 μL of metagenomic DNA, whose concentration is less than 500 ng (*see* **Note 10**).

3. Aliquot 1 μL each of forward and reverse primers so that the final concentration is 0.2 μM (*see* **Notes 20** and **21**).

4. Add 22 μL of nuclease-free water.

5. Mix the contents of the tube and centrifuge briefly to bring all the components to the bottom of the tube.

6. Place the tubes in a thermal cycler and set up the following cycling program:

94 °C, 5 min (Initial denaturation),

Followed by 34 cycles of

94 °C, 0.5 min.

Ta, 0.5 min (*see* **Note 22**).

72 °C, 1 min/kb of DNA.

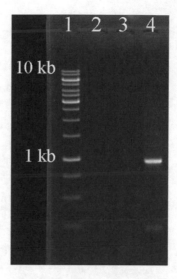

Fig. 6 Agarose gel electrophoresis of asparginase gene PCR amplicons. *Lane 1*: GeneRuler 1-kb DNA Ladder, *Lane 4*: Asparginase PCR amplicon

72 °C, 10 min (Final extension).

Ta: The annealing temperature to be set is characteristic of the primers used (*see* **Note 22**)

7. Analyze the PCR products on 1.2% agarose gel as described in Subheading 3.2. (*see* **Note 23**) (Fig. 6).

8. Recover PCR product of correct size from low melting point agarose gel using the Silica bead DNA gel extraction kit, following the manufacturer's instructions (*see* **Note 24**).

9. Clone the PCR product into pGEM-T Vector as described in Subheading 3.4.

10. Isolate the plasmids from white clones using QIAprep Spin Miniprep Kit (Qiagen), following the manufacturer's instructions.

11. Perform vector-specific PCR as described below to obtain full-length sequences.

3.6 Vector-Specific PCR

1. Set up the master mix as described in Subheading 3.3. Use vector-specific primers such as M13 forward and reverse primers, or SP6 and T7 primers, as detailed in the Subheading 2.

2. Add 1 μL of plasmid DNA (50 ng) as the template.

3. Place the tubes in a thermal cycler and set up the following cycling program:

94 °C, 3 min (Initial denaturation),

Followed by 34 cycles of

94 °C, 0.5 min.

50 °C, 0.5 min.

72 °C, 3 min.

72 °C, 10 min (Final extension)

4. Analyze the PCR products on 1.2% agarose gel as described in Subheading 3.2.

5. Sequence the amplicons by using Sanger's dideoxy method.

6. Determine the identity of the sequences using nucleotide BLAST of the NCBI database.

3.7 Metagenomic Library Screening by Colony PCR

1. Prepare a master mix consisting of 5 µL of 10× PCR buffer, 1 µL of 25 mM dNTP mix, 1 µL each of 0.2 µM forward and reverse primers (*see* **Notes 20** and **21**), 3 µL of 25 mM MgCl$_2$, and 0.1 U of Taq polymerase. Make up the volume to 50 µL using sterile nuclease-free water.

2. Aliquot the master mix to a new labeled microtube. Pick a small amount of white colony and pipette up and down to mix. Mix sufficiently to ensure adequate lysis during the heating steps (*see* **Note 25**).

3. Place the tubes in a thermal cycler and set up the following cycling program:
   95 °C, 5 min (Initial denaturation),

   Followed by 34 cycles of

   95 °C, 1 min.

   Ta, 1.5 min (*see* **Note 22**).

   72 °C, 1 min/kb of DNA.

   72 °C, 10 min (Final extension)

4. Analyze the PCR products on 1.2% agarose gel as described in Subheading 3.2. Determine the insert size by comparing it with the DNA marker.

4   Notes

1. For metagenomic DNA isolation, the amount of starting material should not exceed that mentioned in the protocol, as it can also increase the concentration of inhibitors isolated along with DNA, thereby making PCR and subsequent genetic manipulation difficult.

2. For metagenomic DNA isolation from tissue samples, contamination with host DNA is a major problem. Hence, do not grind such samples. Soil samples, on the other hand, may be vortexed thoroughly and long enough to disperse the particles and to remove the associated organisms.

3. Remove fat and other visible impurities during the process of dissection or handling, prior to suspension of the sample in the extraction buffer. This will prevent clogging of filters.

4. The use of commercial kits such as QIAamp DNA Stool Mini Kit (Qiagen) or PowerSoil DNA isolation kit (MO BIO) for DNA isolation ensures the good quality of DNA, but, according to our experience, there may be a severe compromise in the yield of DNA obtained. This may not pose a problem for PCR, but the lysis method may be biased toward some groups of bacteria. Harsh lysis methods employed in the conventional phenol–chloroform method are recommended in such cases.

5. DNA can be quantified using different techniques. UV spectrophotometry is the most cheap and widely used method. NanoDrop series of spectrophotometers make quantification of DNA using just 1 μL of the sample, when the concentration is in nanogram range. However, they tend to overestimate DNA concentration in the presence of free nucleotides, single stranded DNA, RNA, etc. This can be tested by agarose gel electrophoresis of the DNA and comparison of the bands with standards of known concentration. Fluorometers (such as Qubit 2.0 fluorometer) on the other hand, provide accurate information with even picogram quantities of DNA, as they are based on the intercalation of fluorescence based dyes. Bioanalyzer from Agilent Technologies is a chip-based nucleic acid analysis system that provides accurate quantification with picogram quantities of DNA and provides information regarding the size range of DNA as well. The latter two require the use of several costly reagents and accessories. Choice may be made after careful analysis of cost–benefit ratios.

6. In case a traditional spectrophotometer is used, the following formula can be used to calculate the DNA concentration:

    For 1 cm path length, Optical density at 260 nm (OD_{260}) of 1.0 corresponds to 50 μg/mL of double stranded DNA.

    Hence, Concentration of DNA = 50 μg/mL × OD_{260} × dilution factor.

7. Purity of metagenomic DNA is expressed in terms of absorbance ratios at 260/280 nm, indicative of protein contamination, and 260/230 nm, indicative of organic contaminants. Values of 1.8 and 2 are considered optimal for 260/280 and 260/230 ratios. According to our experience, it is more important to have an ideal 260/230 ratio. Presence of organic contamination inhibits PCR and such samples need to be taken for any analysis only after purification of DNA to the ideal 260/230 ratio.

8. The 16S rRNA gene is highly conserved across bacterial and archaeal species and is considered as the gold standard for bac-

terial diversity analysis. The entire 1.5 kb gene contains alternating conserved and hypervariable regions. The latter act as species specific signatures, permitting easy identification of bacteria. Primer pairs spanning one or more hypervariable regions are used for metagenomic analyses. Presence of 1.5 kb amplicon is also indicative of the presence of bacterial DNA and good quality of the sample. However, the size of the amplicon may vary depending on the primer pair that is used.

9. Next generation sequencing platforms are now widely used to establish the phylogenetic diversity of metagenomes. Short oligonucleotide primers are designed to amplify the hypervariable regions of 16S rRNA gene and bioinformatics tools such as QIIME can be used for taxonomic profiling. This method is cost-effective and rapid, but demands powerful computational resources.

10. It is best to keep a positive control and negative control for all PCR reactions to prevent misinterpretation of positive or negative results.

11. When using insert sizes other than the one specified here, it is important to calculate the molar ratio of PCR product andvector. 3:1 is the preferred ratio in most cases. Higher ratios may also be tried.

12. When cloning larger PCR products, it is advisable to increase the concentration of the insert and ligation time.

13. When TA cloning is involved, it is important to keep the magnesium concentration under check as excess magnesium ion concentration will increase the number of nonspecific bands and, hence, following PCR, the reaction mix should be subjected to gel elution to clone the correct insert.

14. TA cloning can be done into any *E coli* cell that contains the lacZΔM15 mutation, facilitating the identification of recombinants by blue white screening. The vector contains the N-terminal α fragment of the β- galactosidase gene and hosts must express the C-terminal ω- fragment. Production of both fragments results in the restoration of β-galactosidase activity, which cleaves the substrate 5-bromo-4-chloro-3-indoxyl-β-D-galactopyranoside (X-Gal) to form a blue-colored product, in case of cells harboring the empty vector. As the insertion of foreign DNA occurs into the multiple cloning site within the Lac Z-α fragment, β-galactosidase activity will not get restored and, hence, recombinants appear white. The hosts suitable for this cloning include *E coli* JM109, DH5α, DH10B, XL1-Blue, and ER1727.

15. Transformation of the insert ligated vector can be done by several methods. Both electroporation and chemical transformation methods may be employed. The use of kits helps to

save time. In fact, electroporation may give higher transformation efficiencies in certain cases.

16. Plasmid isolation from the phylogenetic library may be done using QIAprep Spin Miniprep Kit (Qiagen), GenElute Plasmid Miniprep Kit (Sigma-Aldrich), or other commercial plasmid isolation kits or alkaline lysis method [12].

17. Amplified Ribosomal DNA Restriction Analysis (ARDRA) is one among the several genetic fingerprinting techniques used to make a microbial community profile based on PCR amplicons obtained from environmental DNA samples. It relies on DNA sequence variations in PCR-amplified ribosomal RNA genes such as the 16S rRNA gene. The amplified DNA is restriction digested with tetracutter restriction endonucleases and the resulting fragments are resolved on agarose or polyacrylamide gels. This does not provide any information regarding the type of microorganisms present in the sample, but can be used for rapid monitoring of microbial community diversity with regard to time or changing environmental parameters. Here, it is used to identify the unique clones and hence unique operational taxonomic units (OTUs). The other fingerprinting techniques used are DDGE/TTGE, RAPD, SSCP, T-RFLP, RISA, LH-PCR, etc.

18. The use of master mixes greatly simplify the detection of specific genes in a metagenome sample, as it will only require the optimization of annealing temperature, which is accomplished by a simple gradient PCR.

19. If PCR amplification is intended for cloning and expression of specific genes, Taq polymerase is not recommended, as its fidelity is less. For such applications, high-fidelity enzymes such as Pfu DNA polymerase, Vent DNA polymerase, and Phusion DNA polymerase are recommended.

20. Primer sequences may be obtained from the literature based on the source of metagenome sample. Otherwise, degenerate primers may be designed based on all the available amino acid sequences in the GenBank database (http://www.ncbi.nlm.nih.gov/genbank/). Primer designing can be done by multiple sequence alignment of all the sequences using ClustalW program and identification of conserved regions. Alternatively, tools such as CODEHOP (http://blocks.fhcrc.org/codehop.html) or HYDEN (http://acgt.cs.tau.ac.il/hyden/HYDEN.htm) may be used for the design of such degenerate primers. Input formats for such tools may differ. For instance, CODEHOP requires multiple sequence alignment of proteins and HYDEN requires input file of DNA sequences. The properties of primers designed may be checked using software or online tools such as OligoAnalyzer 3.1 (IDT) (https://eu.

idtdna.com/calc/analyzer) or Multiple Primer Analyzer (ThermoScientific) (https://www.thermofisher.com/in/en/home/brands/thermo-scientific/molecular-biology/molecular-biology-learning-center/molecular-biology-resource-library/thermo-scientific-web-tools/multiple-primer-analyzer.html) or OligoCalc (http://biotools.nubic.northwestern.edu/OligoCalc.html).

21. Specific primers can be designed if the sequence of the gene to be amplified is already known, when whole metagenome sequencing has been done. This can be done either manually or by the use of tools such as Primer3 (http://bioinfo.ut.ee/primer3/), Primer BLAST (http://www.ncbi.nlm.nih.gov/tools/primer-blast/), and PrimerQuest (http://eu.idtdna.com/primerquest/home/index).

22. It is imperative to give the correct annealing temperature for PCR reaction to occur. The annealing temperature is 5 °C lower than the melting temperature of the primers. The melting temperature is calculated using the following formula:

$$Tm\left(^{\circ}C\right) = 4\left(NG + NC\right) + 2\left(NA + NT\right)$$

where NG, NC, NA, and NT are the numbers of guanine, cytosine, adenine, and thymine bases in the primer.

Though the formula predicts the annealing temperature, it is always advisable to keep a gradient PCR as the sequence of the genes in the sample is unknown. Annealing temperatures must be predicted Ta ± 5 °C.

23. Some PCR master mixes such as the EmeraldAmp GT PCR Master Mix contain the loading dye and hence need not be mixed with loading dye prior to loading onto agarose gel.

24. Several gel extraction kits are available such as the Silica bead DNA gel extraction kit (Thermoscientific), QIAquik gel extraction kit (Qiagen), GenElute Gel Extraction Kit (Sigma-Aldrich). If the gene specific PCR reaction gives a single band on the gel electrophoresis, it needs not to be gel purified. It can be applied to purification directly, using the silica bead purification kit, following the protocol for DNA purification from reaction mixture or as the kit used specifies.

25. The amount of colony picked for PCR is very crucial, as it may give false positives as well as false negatives. When more number of cells are picked, they tend to inhibit the PCR reaction. Barely touching the bacterial clone will suffice. So, to improve the accuracy of the process, the bacterial colonies picked may be first pipetted into a fresh PCR tube with sterile water and boiled at 95 °C for 15 min for cell lysis. 1 μL of this may be used as a template.

References

1. Handelsman J (2004) Metagenomics: application of genomics to uncultured microorganisms. Microbiol Mol Biol Rev 68:669–685

2. Miller DN (2001) Evaluation of gel filtration resins for the removal of PCR-inhibitory substances from soil and sediments. J Microbiol Methods 44:49–58

3. Tina KJ, Bindiya ES, Raghul Subin S, Bhat SG (2014) Appraisal of extraction protocols for metagenomic DNA from fish gut microbiota. IJAIR 3:7–13

4. Devi SG, Fathima AA, Radha S, Arunraj R, Curtis WR, Ramya M (2015) A rapid and economical method for efficient DNA extraction from diverse soils suitable for metagenomic applications. PLoS One 10:e0132441

5. Gutiérrez-Lucas LR, Montor-Antonio JJ, Cortés-López NG, del Moral S (2014) Strategies for the extraction, purification and amplification of metagenomic DNA from soil growing sugarcane. Adv Biol Chem 4:281–289

6. Woese CR (1987) Bacterial evolution. Microbiol Rev 51:221–271

7. Hamady M, Knight R (2009) Microbial community profiling for human microbiome projects: tools, techniques, and challenges. Genome Res 19:1141–1152

8. Kotik M (2009) Novel genes retrieved from environmental DNA by polymerase chain reaction: current genome-walking techniques for future metagenome applications. J Biotechnol 144:75–82

9. Itoh N, Isotani K, Makino Y, Kato M, Kitayama K, Ishimota T (2014) PCR-based amplification and heterologous expression of pseudomonas alcohol dehydrogenase genes from the soil metagenome for biocatalysis. Enzyme MicrobTechnol 55:140–150

10. Shivaji S, Bhanu NV, Aggarwal RK (2000) Identification of *Yersinia pestis* as the causative organism of plague in India as determined by 16S rDNA sequencing and RAPD-based genomic fingerprinting. FEMS Microbiol Lett 189:247–252

11. Yaacob MA, Hasan WA, Ali MS, Rahman RN, Salleh AB, Basri M, Leow TC (2014) Characterisation and molecular dynamic simulations of J15 asparaginase from *Photobacterium* sp. strain J15. Acta Biochim Pol 61:745–752

12. Sambrook J, Russell DW (2001) Molecular cloning: a laboratory manual. Cold Spring Harbor Laboratory Press, Cold Spring Harbor

Chapter 18

Arbitrarily Primed PCR for Comparison of Meta Genomes and Extracting Useful Loci from Them

Leigh Burgoyne, Lin Y. Koh, and David Catcheside

Abstract

A method is described that uses arbitrarily primed PCR followed by many cycles of amplification under stringent conditions and selection by computational means to obtain a set of sequence tags that can be used for the comparison of metagenomes. Relative to unselective shot-gun sequencing, the results are small data sets that can be csompared electronically or plotted as scattergrams that are simple to interpret. The method can be used to compare groups of samples of any size to build in-house databases from which, for example, the provenance of trace soil samples may be inferred. The method also allows for selection of primers with locus-specificity and an example is given in which a South Australian sequence-related to a Portuguese thermophile (*Rubrobacter radiotolerans*) is extracted and tested on a set of soils.

Key words Metagenome analysis, Soils, Forensic, PCR, 50mers, Sequence tags, Primer design

---

1 Introduction

The development of methods for metagenome analysis is a dynamic field [1–5] which faces the problem that, in many cases, a small subset of species dominate the DNA present, which are not necessarily the most useful for discriminating between metagenomes. This is of particular importance for discrimination between samples in analysis of soils and detritus for example for forensic purposes. In consequence, to obtain an estimate of species diversity for discrimination between samples, it is common to sequence the metagenome DNA either in considerable depth or selectively. The former is costly and the computational needs to process and store the voluminous data for comparative purposes are substantial. Selective amplification on the other hand poses the problem that the data generated only include sequences from the range of cellular organisms and viruses that the primer set can amplify, which seriously prejudges the outcome before even beginning a study. A method is described here which substantially overcomes these difficulties by identifying a set of highly discriminating 50mer

Lucília Domingues (ed.), *PCR: Methods and Protocols*, Methods in Molecular Biology, vol. 1620,
DOI 10.1007/978-1-4939-7060-5_18, © Springer Science+Business Media LLC 2017

sequence-tags selected from any whole metagenome by computation following arbitrarily primed long-cycle PCR. This gives compact data sets and does not prejudge the outcome.

A single 18mer primer is used to amplify DNA extracted from a metagenome. The amplification uses low stringency to pick up sequences distributed across the entire DNA present in the metagenome and then uses high stringency for a large number of cycles to selectively amplify a subset of sequences. This process exploits an intrinsic bias in PCR amplification which is dependent on properties of each amplicon [6]. These properties are dictated by the base sequence of the amplimer primarily when it is single-stranded during the annealing phase of amplification. Thus each sequence has a unique efficiency with which it is amplified relative to the other sequences being coamplified with it. This inherent filtering of the amplification leads to a progressive bias in the composition of the population of amplicons that favors a specific subset of those present in each amplification cycle following the low stringency initial cycles of amplification. No single amplicon takes over due to a number of interacting factors, the most important of these being interference in primer loading due to homology of the termini and interference with primer loading due to renaturation of those amplimers present in higher frequency. These and other factors lead in practice to repeatable subsets of sequences being present following random priming, and high stringency, high cycle number, amplification forming a signature for any given sample [6, 7]. This chapter describes one set of conditions for such analyses.

The sequences that predominate after amplification are self-selected by their structural and amplification properties, not their levels in the template, and so are not necessarily dominated by sequences derived from the high-frequency species that are contributing the most DNA to the metagenome. In theory, an expected result of this is that, per Mb of DNA sequenced, a greater diversity of the sequences offered by the original metagenome is displayed in the output. In practice, as few as 3000 reads, each of approximately 250 bases, can suffice to discriminate between metagenomes sampled in this way, though more will usually be preferred, so multiplexing with sequence tags allows a single batch of a mass parallel sequencing to suffice for multiple samples, with the realities of sequence tagging setting the practical upper limit.

To compare mixed genomes or metagenome samples, the data is reduced in size by mining it for 50mer sequences that have sufficient complexity to be discriminating between metagenomes. The length 50 was chosen as a useful basis for comparison of the oligomers, since with this length the number of unique identifiers possible is gigantic (4 to the 50th power divided by two $\sim 6.3 \times 10^{29}$). The most informative sequences for discriminating between samples, such as between the soils used in the examples given here, are those of high sequence complexity which are devoid of homopolymer runs and short sequence repeats which are commonly found in some taxons. The process of mining the sequence data for discriminating 50mers uses a set of rules (*see* Subheading 3) that only accepts 50mers which have characteristics of genes or pseudogenes or similar, where the sequence approximates to a random sequence. The selection algorithm uses the sum of square deviation of dinucleotide frequency from that of a truly random DNA sequence having equal numbers of each of the four bases. The selection process also limits the number of 50mers that are allowed from any amplimer to avoid giving preference to long amplimers. Sequences that pass these filters are binned and the collection of bins and their contents comprises a signature of the genome or metagenome from which they were derived, as detailed in Subheading 3.

For each of the separate metagenomes, the list of the discriminating 50mers present and the number of times each was observed is taken as the identifier for that metagenome. Those data can be compared in silico, and pairwise comparisons can usefully be visualized by two dimensional scatter-plots in which the number of occurrences of each of the discriminating sequences in the metagenomes being compared gives the coordinates of a data point in the plot. Where there are no shared 50mers, all data points are on one or other axis, as in Fig. 1, but if there are shared 50mers those data

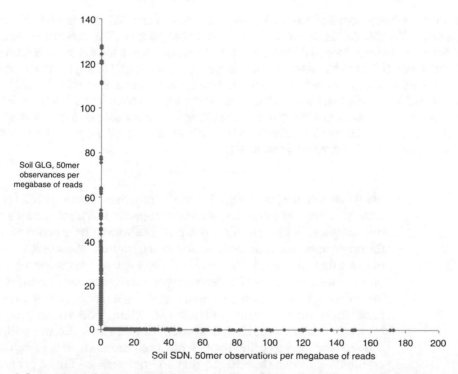

Fig. 1 Comparison of two soil metagenomes that show no relationship. Suburban nature-strip soils 6.8 km apart, GLG at S34.97441 E138.51410 and SDN at S35.032489 E138.539134 are compared in a scattergram. Each *data point* on the scattergram represents the frequency of occurrence of a specific 50mer in the two metagenomes. All the data points are on an axis; the two soils do not share any 50mers

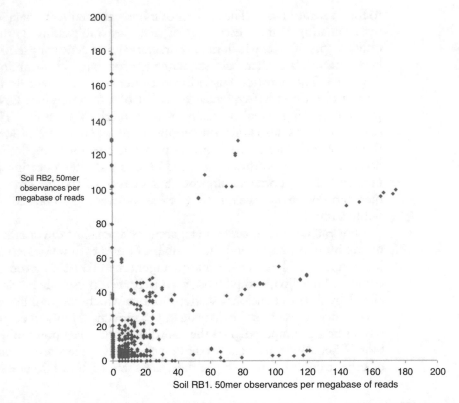

Fig. 2 Comparison of the metagenomes of two related soils. Samples RB1 and RB2 were taken 6 m apart along a vineyard-row at S34.59558 E138.882081 that had recently been ploughed. The occurrence of many 50mers shared by the two samples, those data points that are off axis, shows that the two soils are related. The nonrandom distribution of the shared data-points close to the origin is an artefact arising from the digitizing effect of small numbers. However, clusters of data points around sloping lines in the body of the scattergram reflect 50mers that are in definite ratios to each other in the two metagenomes. This effect arises from at least three sources. First, long reads donating more than one 50mer to the array, second, 50mers that arise from the same species and third, 50mers that have arisen from different species but are present in definite ratios to each other due to their soils' community structure

points are off axis, as in Fig. 2. Such comparisons can be used for a variety of forensic purposes, such as comparison of trace detritus or soil samples for determining the possible source of a proband or for environmental monitoring. Soil metagenomes show a decreasing number of shared 50mers with distance of separation of the sample sites, and soils of different types often have no discriminating 50mers in common [8], making the detection of trace contamination from another soil feasible. Here, one strategy is to return to the raw sequence data used for extracting discriminating 50mers and design primer sets from the parent sequences of discriminating 50mers specific to that soil or detritus. Those primers would then allow detection of trace amounts of the target soil or detritus, for example inferring that a person has been at a place relevant to a criminal case.

Although the selected-50mer approach for metagenome analysis does not set out to identify the species from which the discriminating 50mers are sourced, BLASTn searches of the rapidly expanding GenBank database and the growing assemblage of whole genome data bases may identify the genera of origin if not the species.

Here we present protocols for arbitrarily primed high-cycle number PCR to extract profiles of representative amplicons from a metagenome, a way to extract discriminating 50mers from the amplicon sequences, and ways to compare and interpret the data sets obtained.

2   Materials

Solutions are prepared using analytical reagents, deionized MilliQ-water, and nuclease-free water for PCR by sterilization for 30 min in a UV chamber.

2.1   Soil DNA Preparation

1. AR grade isopropanol.
2. Acid-treated and neutralized silica sand for preparation of DNA from control samples (e.g., from chicken tissue).
3. MOBIO PowerSoil^R DNA isolation kit for DNA extraction.
4. Soil DNA extraction kits and PCR purification columns are stored at room temperature and all PCR reagents at −20 °C or lower.
5. NanoDrop™ spectrophotometer.

2.2   PCR Amplification

1. Primers from Geneworks Pty Ltd. South Australia.
2. The primer for arbitrary priming is: GGAGGTGGTG TTCGAGGG (designated antiseq05).
3. Promega GoTaq Flexi DNA polymerase and Promega deoxy-nucleotides (dNTPs) for PCR.
4. AdBiotech™ PCR purification columns to purify PCR products.

2.3   PCR Product-Monitoring Gels

1. BioRad certified molecular biology agarose.
2. GelRed™ nucleic acid stain for agarose gel.
3. TAE electrophoresis buffer, 50× stock: Dissolve 242 g of Tris base in 750 mL deionized MilliQ water and 14.6 g EDTA (acid) in 100 mL deionized MilliQ water. Mix and add 5.71 mL glacial acetic acid and adjust pH to 8.3. Store at room temperature. Dilute 1 in 50 for use.
4. Loading dye (6×): Dissolve 0.025 g bromophenol blue and 0.025 g xylene cyanol in 3 mL glycerol. Make up to 10 mL with deionized MilliQ-water.
5. For preparation of 2% agarose with GelRed™. Add 1.2 g molecular grade agarose to 60 mL TAE. Melt agarose gradually in a microwave oven until completely dissolved. Cool for

Table 1

26mer sequence-coding primers suitable for terminal labelling of PCR fragments of primer antiseq05. The 12 bases at the 5′ end provide the code, the remaining 14 are homologous with primer antiseq05

5′ AATTTAATGTAAGTGGTGTTCGAGGG
5′ TTTAATATTGATGTGGTGTTCGAGGG
5′ TTTAATATGTATGTGGTGTTCGAGGG
5′ AAATTATATGTAGTGGTGTTCGAGGG
5′ AAATTATAGTTAGTGGTGTTCGAGGG
5′ TTTAATTAGATAGTGGTGTTCGAGGG
5′ TTTAATTATAGAGTGGTGTTCGAGGG
5′ TTTAATAATGAAGTGGTGTTCGAGGG
5′ AAATTATTATGAGTGGTGTTCGAGGG

10 min at room temperature, add 6 μL GelRed™, mix gently and pour onto a 8.2 cm by 12.5 cm plastic tray with a plastic comb (teeth 5 mm wide, 1 mm deep). Leave to set completely for 20 min. Remove plastic comb carefully ensuring wells are not damaged. Submerge agarose in TAE and electrophorese at 1 volt per cm of the gel plate for 90 min (*see* **Note 1**).

2.4 Sequence-Coding PCR Fragments Termini

Table 1 lists eight 26mer sequence-coding primers designed to allow multiple samples to be processed together. These primers are partially homologous to the original primer and, partially, a distinctive label. They have been designed to minimize amplification distortions due to the presence of the sequence-coding tail and to minimize misidentification by having at least three nonessential bases at the 5′ end to provide for bases that may be lost during processing for massed parallel sequencing.

3 Methods

3.1 Extraction of Metagenome DNA from Soils

1. Field collection: 2–5 g of soil is taken from the upper zone of the soil, typically 0–2 cm deep, placed into an excess of isopropanol (≥1/10 w/v) as a dehydrating agent and preservative during storage.

2. Separate soils from isopropanol by centrifugation and air-dry or vacuum-dry at room temperature.

3. Extract DNA from a 0.25-g sample of dried soil using a commercially available soil DNA extraction kit (for example MOBIO PowerSoil^R DNA isolation kit) according to the manufacturer's instructions.

4. Measure DNA concentrations with a NanoDrop™ spectrophotometer and dilute soil DNA to 5 ng/μL with sterile UV irradiated water.

3.2 Preparation of Reference DNA from Chicken

1. Approximately 5 mg of chicken flesh is ground with 0.25 g of dry, acid-treated, and neutralized sand. This is then isopropanol-dried and DNA-extracted as if for a soil.

3.3 PCR Amplification of Metagenome DNA Template

Since amplifications are vulnerable to contamination with amplified products from previous experiments, the assembly of PCR mixes should be rigidly separated from processing after the first round of PCR.

The protocol preferred here uses two rounds of low stringency amplification to acquire a set of sequences broadly distributed across the template DNA. Other acquisition conditions have been explored (see **Note 2**).

1. In the preferred protocol, the first round PCR reaction contains 5 ng soil DNA template in a reaction volume of 50 µL, 0.6 µM primer anti-seq05, 1x "colorless GoTaq" reaction buffer, 2 mM $MgCl_2$, 0.2 mM dNTP mix, and 1.25 units GoTaq Flexi DNA polymerase.

2. In the examples given here, the conditions for the first two rounds of amplification were: 94 °C for 5 min, 2 low-stringency (acquisition) cycles of (94 °C for 30 s, 30 °C for 3 min, 72 °C for 3 min). This was followed by 35 cycles of (94 °C for 30 s, 62 °C for 30 s, 72 °C for 3 min) and a final extension of 72 °C for 7 min.

3. Reproduction of the outcome of these conditions is best done with test amplifications (see **Note 3**), which should be monitored by gel electrophoresis (see **Note 4**). For estimation of the degree of coverage by this procedure see **Note 4**.

4. The conditions used here and subsequent processing of data are a compromise between discrimination power and size of the database used for comparison of samples. Databases to be used by more than one laboratory require adoption of the method as described here or an agreed variant (see **Note 5**).

3.4 Sequence-Coding Amplification

1. The 50 µL PCR reaction mix contains 3 µL of the first round PCR product as template, 0.6 µM truncated anti-seq05 tagged primer (Table 1), 1x "colorless GoTaq" reaction buffer, 2 mM $MgCl_2$, 0.2 mM dNTP mix, and 1.25 units GoTaq Flexi DNA polymerase.

2. Amplification conditions: 94 °C for 5 min, 2 cycles of (94 °C for 30 s, 51 °C for 30 s, 72 °C for 3 min) followed by 35 cycles of (94 °C for 30 s, 62 °C for 30 s, 72 °C for 3 min) and a final extension of 72 °C for 7 min (see **Note 6**).

3.5 Massed Parallel Sequencing of Amplicons

1. Pool equal volumes from second round amplifications, purify using an AdBiotech PCR purification column, or similar, following instructions provided by the supplier and elute with sterile nuclease-free water.

2. Measure DNA concentration using a NanoDrop™ spectrophotometer. For the examples given here, 400-ng samples in 20 µL water were sequenced using ion torrent™ by the Australian Genome Research Facility (*see* **Note 7**).

3.6   Data Processing

The methods described here are specifically applicable to data generated by arbitrary priming and are not well adapted or useful with most data from shotgun sequencing.

1. On receipt of raw data from a massed parallel sequencing technology, it is first converted into text and the quality information discarded.

2. Any characters other than A, T, G, or C are interpreted as break points in a sequence and the resultant fragments become separate sequences.

3. Primer sequences are removed from the termini of all sequences which are also scanned for internal sequences related to the primers, with relatedness being defined as clusters of hexamers that match the primer or its complement. Clusters of this sort can be derived from the primer and are excised, and the flanking regions treated as separate sequences.

4. Any sequences produced by these processes that are shorter than 50 bp are discarded and the purged residuum is regarded as the "useful reads." This population of sequences should be archived against any future need for further data mining, such as extracting locus-specific primers.

5. All possible 50mers are extracted from a read, then the informational value of each 50mer is taken to be the converse of its SSQD, which is defined as the *sum of squares deviation from random-expectation* for the content of the dinucleotides in the 50mer concerned. From each read, only the 50mer of highest informational value, i.e., with the lowest SSQD, is stored and then only if the SSQD is less than a user set threshold. For the examples included here, a value of 70 was used.

6. At the finish of collecting acceptable 50mers, the number of times that each 50mer was observed amongst the reads is recorded and those observed only once are discarded. The data are stored as a two column array in which each acceptable 50mer is associated with the number of times it was seen, normalized to per megabase of read for reads from a single soil or other metagenome source. Collections of such arrays form a database from which comparisons of metagenomes are made.

3.7   Data Interpretation

To compare two metagenomes, their two-column associative arrays are combined into three columns; one column lists the 50mer sequences and the other two list the occurrences per megabase of reads observed for each metagenome. One line of such an array might be:

AATTTAGGCCCAAATACGGCAGAGGATATTCAATG
GCATTAGCTGATCTT, 190.28, 264.58.

The pairs of scores are then used as the coordinates of a data point in a scattergram, as shown in Figs. 1 and 2. As the associative arrays typically contain only a few thousand lines, the data can easily be held within and plotted by spreadsheets such as Excel.

Figures 1 and 2 show scatterplots derived from comparison of datasets obtained from the metagenomes of unrelated and closely related soils. The soils compared in Fig. 1 are both samples of urban roadside soils taken from sites some kilometers apart. They show no similarity as all 50mers are found either in one soil or the other; none of the data points are off-axis. This illustrates the discrimination potential of the method. The soil samples compared in Fig. 2 are expected to be closely related as they are taken from locations separated by a few meters. As expected, they share many 50mers, shown by data points off-axis, yet they also have 50mers that are unique to each sample, those data points that are on-axis, showing the potential of the method for detection of short range variation in soil metagenomes related to, for example, contamination of a soil or other metagenome with toxic compounds or burial of organic materials.

For fast operation of a database query, the original associative arrays are all that is required. For the highest quality comparisons, all the 50mers concerned in both probands' arrays are best rescored against each proband's reads used to generate the two original associative arrays; this is because random and systemic variations between the lengths of reads from different sequencing runs may lead to different choices of the best 50mers from even identical populations of PCR products. This effect can be dealt with by extracting the 50mers from full length PCR products assembled by contiging before scoring them against reads or, more simply, by rescoring both sets of reads against the combined 50 mers from both sets as described. In the examples given here, the latter method was used.

3.8 Extraction of Targeted Primers for Specific, User-Chosen Applications

While the data sets from arbitrarily primed PCR are very small compared to those from shotgun sequencing of similar utility, they are sufficiently complex to be mined for targeted primers.

A 50mer occurring at high frequency in a metagenome that is of interest can be used to extract parent reads from the archived "useful reads." Due to the amplification conditions, unlike shotgun sequencing, such high frequency reads will not necessarily come from the most common organisms in the metagenome and the high read numbers eliminate sequencing errors. Polymorphic regions may be included or excluded according to the user's application.

3.8.1   An Example: Designing Primers for a Sequence with the Potential to Be a Region-Specific Marker

A 50mer present in high copy number in the metagenome of an unremarkable suburban soil (SDN), an un-irrigated red brown loam from a roadside verge at S35.032489 E138.539134, was investigated. The parent reads for the 50mer were extracted from the archived "useful reads," aligned and the consensus sequence used to search NCBI databases using blastn. The single nontrivial BLASTn match (77% over 465 bases) was to *Rubrobacter radiotolerans* strain RSPS-4 complete genome [9] from hot springs in Portugal. Homology to a northern hemisphere hot spring extremophile in DNA from a southern hemisphere soil metagenome that experiences hot summers suggested potential for those sequences to be a region-specific marker. Primer sets were designed using Primer-BLAST (NCBI) yielding forward primer SDNf1 and reverse primer SDNr1 for a single product 262 bases long (Table 2).

In a pilot study, 0.4 µM SDNf1 and SDNr1 were used in a 50 µL reaction mix containing 5 ng soil DNA as template, 1× "colorless GoTaq" reaction buffer, 2.0 mM $MgCl_2$, 0.2 mM dNTP mix, and 1.25 units GoTaq Flexi DNA polymerase. Amplification conditions: 94 °C 5 min, followed by 35 cycles of (94 °C 30 s, 62.5 °C 30 s, 72 °C 3 min) and a final extension of 72 °C 7 min.

Of 15 soils randomly selected from across South Australia, eight yielded products all of the correct size (Table 3). None of the other positive soils (Table 3) came from near the soil of origin. Although such a small sample of soils is an insecure basis for solid conclusions, it is intriguing that all the positive soils were either from semi-arid or arid or unwatered (as was the soil of origin) locations. However, some arid soils were negative so it was not an invariant marker for aridity, even within in the state of South Australia, and many more samples would be required before such observations had any utility. Nevertheless, this is an example of a

Table 2

Consensus sequence from soil SDN that provided the 50mer with homology to *Rubrobacter radiotolerans*. The consensus sequence was assembled by aligning primary reads having homology to the 50mer and reads overlapping those reads. The SDN-derived primer pair SDNf1 CAGGTACGGCTCGTTCTTGT and SDNr1 TCGCTACGAGTTCCAGATGC is shown by *underscores* and the 50mer in *bold*

AGAGGACGATCTCCCCGTCCGTCTCCACCCGCTCCCCGCCGATGATTAGGGGGTAGCTC
CTGCCAAGCGAGCCCTCTACCTCTTCCAGGGCTTTCCGCATGGCCTTCC**GGTTCGAC
TCGTCCGTCCAGTC** <u>CAGGTACGGCTCGTTCTTGT</u> ACGGCAACAGTCCCATAATCT
TCCTCCTGCGAGCGGCTACCGTCTGACGGTGTTCAAGATGCTCTCAATGACGTCTCTG
GCCACGTAGCTGGCGATCTTCGGGTTCTCCTTTAACCTGCGCGTGGAGTACTCGTACC
AGTCCTTGCCGAACGGCACGTATACCCTCACCTTGTGCCCCGCATCGACGAGTATACG
CCGTAGTTCTTCGTCGACCCCAAGCA<u>GCATCTGGAACTCGTAGCGAT</u>CCCTGGACAAA
CCCATCGGTGGATCAAGCGCAGACCGTGCCAGACGAGATATTCGTCGTGGGTAGCGA
TCCCTACGTACG

Table 3

Localities tested for product from primers SDNf1 and SDNr1. Yes: a product close to the expected length of 262 bases was observed by gel electrophoresis. No: no product of any length was observed

Site code	Location	Geodata	Product
SDN	Seaview Downs suburb	S35.032489 E138.539134	Yes
GWA	Goolwa	S35.50414 E138.78529	No
NAR	Naracoorte caves	S37.036025 E140.796641	No
MTG	Mt Gambier	S37.83240 E140.76819	No
AGV	Angle Vale district	S34.62875 E138.68077	No
VIR	Virginia	S34.64843 E138.55446	Yes
JCR	Jacob's Creek district	S34.59931 E138.90245	Yes
NPV	Nurioopta district	S34.46408 E138.99876	Yes
DVS	Daveystone district	S34.49755 E138.83559	No
HBY	Highbury district	S34.85194 E138.71095	Yes
DC8	Lake Everard Station	S31.107559 E135.896641	Yes
KNGRY9	Kingoonya	S31.03528 E135.37547	Yes
DM3	Mt Rescue	S35.5623 E140.1536	No
MBCR	Mambray Creek	S32.84140 E138.03015	No
WCP6	Whyalla 6	S32.947475 E137.56368	Yes

widely distributed marker locus defined by a primer set mined out of the arbitrarily primed data that may be a useful component of a region specific diagnostic soil multiplex.

4   Notes

1. The methodology hereby described should be suitable for monitoring PCR products up to approximately 1.5 kb.

2. Another method for acquiring sets of fragments with the same primer have been used [7]: 0.4 μM of the single arbitrary primer with 2.5 mM Mg^{2+}, 0.2 mM of each dNTP, 0.5 U HotstarTaq DNA polymerase (Qiagen), 1× HotstarTaq buffer (Qiagen), and 1–5 ng of the extracted soil DNA as a template in a volume of 25 μL using an amplification regime of 95 °C for 15 min, 42 cycles of 94 °C for 30 s, 55 °C for 30 s, 72 °C for 60 sec, and a final extension of 72 °C for 7 min.

3. High amplification PCR is notoriously difficult to precisely reproduce from published conditions. When using the primer specified here and synchronizing one laboratory with another,

it is expedient to use a large genome such as that of a verte-brate genome (e.g., chicken) as a standard along with a small number of soil DNAs and DNA-free sand (acid-treated and neutralized or heated to a high temperature). The sand blank allows a check for contaminants in reagents and chicken tissue ground in sand and extracted as if a soil provides a positive control. The test samples and soils are then amplified by the protocol described here and electrophoresed on an agarose gel. When conditions are correct, the chicken sample will yield bands that range from 1.5 kb or larger down to approximately 300 bp (Fig. 3) and soil samples usually a smear with some bands amongst them. If necessary, the annealing temperatures should be modulated up or down one or a few degrees until the reaction gives, approximately, the illustrated result, an exact match is not necessary. The degree of smear and bands will differ for different soils, reflecting the degree of sequence diversity. Well defined bands without much background smear is an indication that DNA template concentration may be unac-

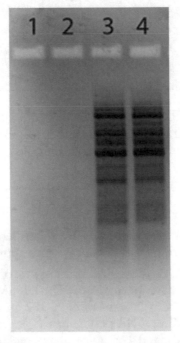

Fig. 3 Specimen gel of amplification products obtained from reference DNA. Chicken tissue prepared as described in Subheading 3.2 was amplified using the preferred protocol described in Subheading 3.3 and electrophoresed in a 2% agarose gel as described in Subheading 2.3. *Lanes 1* and *2* replicate amplifications of sand-only blanks. *Lanes 3* and *4*, replicate amplifications of chicken DNA. The overlapping series of bands range from approximately 300–1500 bp. Control amplifications using chicken DNA or another large genome provide verification that amplification conditions are appropriate (*see* **Note 3**). Soil DNA templates yield unpredictable patterns ranging from banding patterns to a polydisperse smear

ceptably low. Blank amplifications typically give this result when reagents have minute trace DNA contamination. Such contamination rarely interferes in practice as the added template dominates such amplifications.

Until confidence is gained in the reliability of amplifications, DNA from a vertebrate with a known genome should be routinely used as a control, and randomly chosen sequences from the outcome checked against that vertebrate genome. If chicken is used that can be checked at NCBI [http://www.ncbi.nlm.nih.gov/].

4. Given the complete lack of boundaries, either of the species present or their relative contributions to the population of sequences in a metagenome, the degree of coverage can only be inferred indirectly by using exactly the same procedure on a real genome, such as the human genome, or a known mixture of genomes and taking the coverage of that as an objective index.

5. The preferred amplification and processing procedures described here is designed to extract a relatively small number of highly distinctive 50mers, 250–500 loci within a metagenome and will achieve this even from a single large genome. The degree of coverage of the genome depends on the details of the primary amplification and can be varied to provide greater or lesser breadth of coverage as a user may choose. The breadth of coverage is largely determined by the settings used for amplification. At one extreme, the range of products is restricted by high stringency of amplification whilst lowering stringency leads to growing waste due to primer artefacts, principally concatamers; but it also leads to large databases and thus greater storage requirements and longer processing time for comparisons.

In the processing described here, the distinctive 50mers are taken from the termini of the PCR products. However, the method [7] in which the whole length of the PCR products are used without preference for termini appeared to give results at least as useful.

6. Sequence coding can be monitored by gel electrophoresis as it is expected to distort the original gel pattern. Sequence coding of the fragments post shearing for next generation sequencing, rather than just the amplicon termini may be preferred, is quite practical and has been used [7].

7. Ion torrent™ is one of the short fragment sequencing methods. As massed parallel sequencing improves for both speed and convenience, particularly when it becomes inexpensive to read kilobase-sized sequences, whole PCR products will be read from one end to the other rather than fragments of them. This will allow for even more data condensation but require

modification of the data-processing procedures. However, even with short-read technology this further condensation, at least in principle, is possible even now after assembling reads into whole PCR products, but at the expense of more processing time.

References

1. Riesenfeld CS, Schloss PD, Handelsman J (2004) Metagenomics: genomic analysis of microbial communities. Annu Rev Genet 38:525–552.   doi:10.1146/annurev. genet.38.072902.091216

2. Wooley JC, Godzik A, Friedberg I (2010) A primer on metagenomics. PLoS Comput Biol 6(2):e1000667.   doi:10.1371/journal. pcbi.1000667

3. Sharpton TJ (2014) An introduction to the analysis of shotgun metagenomics data. Front Plant Sci 16:209. doi:10.3389/fpls.2014.00209

4. Oulas A, Pavloud C, Polymenakou P, Pavlopoulos GS, Papanikolaou N, Kotoulas G, Arvanitidis C, Iliopoulos I (2015) Metagenomics: tools and insights for analyzing next-generation sequencing data derived from biodiversity studies. Bioinform Biol Insights 9:75–88. doi:10.4137/BBI.S12462

5. Khodakova AS, Smith RJ, Burgoyne L, Abarno D, Linacre A (2014) Random whole metagenomic sequencing for forensic discrimination of soils. PLoS One 9:e104996. doi:10.1371/journal.pone.0104996

6. Waters JM, Eariss G, Yeadon PJ, Kirkbride KP, Burgoyne LA, Catcheside DEA (2012) Arbitrary single primer amplification of trace DNA substrates yields sequence content profiles that are discriminatory and reproducible. Electrophoresis 33:492–498. doi:10.1002/elps.201100359

7. Khodakova AS, Burgoyne L, Abarno D, Linacre A (2013) Forensic analysis of soils using single arbitrarily primed amplification and high throughput sequencing. Forensic Sci Int Genet 4:e39–e40. doi:10.1016/j.fsigss.2013.10.019

8. Burgoyne L, Koh L, Catcheside D (2015) A study of one soil, its relatives and contaminants by arbitrary primed PCR with 50mer based analysis. Forensic Sci Int Genet 5:e503–e505. doi:10.1016/j.fsigss.2015.09.199

9. Egas C, Barroso C, Froufe HJC, Pacheco J, Albuquerque L, da Costa MS (2014) Complete genome sequence of the radiation-resistant bacterium *Rubrobacter radiotolerans* RSPS-4. Stand Genomic Sci 9:1062–1075. doi:10.4056/sigs.5661021

INDEX

Lucília Domingues (ed.), *PCR: Methods and Protocols*, Methods in Molecular Biology, vol. 1620, DOI 10.1007/978-1-4939-7060-5, © Springer Science+Business Media LLC 2017